About Time
Speed, Society, People and the Environment

Edited by Tim Aldrich, Forum for the Future

This book has been funded by the Forum for the Future's Business Futures Fund. A number of the Forum's business partners contribute to this fund to support fresh thinking on sustainability issues.

Forum for the Future's mission is to accelerate the building of a sustainable way of life, taking a positive, solutions-oriented approach. As a charity we use our research, experience and learning techniques to work with decision-makers in key sectors. These include business, finance, government and higher education. For more information see www.forumforthefuture.org.uk

Thanks are due to Vidhya Alakeson, Stella Bland, Hannah Bullock, Stephanie Draper, Roger East, James Goodman, Britt Jorgensen, Mireille Kaiser, Paul Miller, Anne Paintin, Jonathon Porritt, Chris Sherwin, John Stuart, Sally Uren and Martin Wright for their support and advice from the genesis of this book in 2002 to its publication.

This book has been the work of many hands, though particular mention should be made of Britt Jorgensen, whose tenacious enthusiasm brought the project to life and gave it its momentum.

about time

Speed, Society, People and the Environment

Edited by **Tim Aldrich, Forum for the Future**

including contributions from
**Will Hutton, Mary Warnock, Martin Rees, Ghillean Prance,
Jay Griffiths, David Boyle, Geoff Mulgan, Alexandra Jones
and Jonathon Porritt**

Routledge
Taylor & Francis Group

LONDON AND NEW YORK

2 0 0 5

First published 2005 by Greenleaf Publishing Limited

Published 2017 by Routledge
2 Park Square, Milton Park, Abingdon, Oxon OX14 4RN
711 Third Avenue, New York, NY 10017, USA

Routledge is an imprint of the Taylor Francis Group, an informa business

Copyright © 2005 Taylor & Francis

Cover by LaliAbril.com.

British Library Cataloguing in Publication Data:
    A catalogue record for this book is available from the British Library.

ISBN 978-1-874719-91-5 (pbk)

Contents

Introduction

Tim Aldrich and
Mireille Kaiser

From the start of the Industrial Revolution more than 200 years ago, developed nations have achieved ever greater prosperity and higher living standards. But through this period our activities have come to affect our atmosphere, oceans, geology, chemistry and biodiversity.

What is now plain is that the emission of greenhouse gases, associated with industrialisation and strong economic growth from a world population that has increased sixfold in 200 years, is causing global warming at a rate that began as significant, has become alarming and is simply unsustainable in the long term. And by long-term I do not mean centuries ahead. I mean within the lifetime of my children certainly; and possibly within my own. And by unsustainable, I do not mean a phenomenon causing problems of adjustment. I mean a challenge so far-reaching in its impact and irreversible in its destructive power that it alters radically human existence.

Tony Blair, Prime Minister, 14 September 2004[1]

---

1 Speech to the Prince of Wales Business and the Environment Programme on its 10th anniversary.

Sustainable development is about time. The origin of this book came from a sense that by exploring the relationship between time and sustainable development we could inspire, agitate and challenge readers to take different perspectives on the problems the Earth faces in the 21st century. We decided to ask experts to think about different aspects of time: from the long time spans of the cosmos to the instantaneous globe-spanning reaction of electronic systems on the Internet, from environmental degradation to work–life balance, politics to ethics — all issues critical to the long-term sustainability of life on planet Earth.

Before we launch into an exploration of perspectives on time and the cosmos we would like to suggest some thoughts you might wish to bear in mind, to use them as provocations as you read.

Taking time

Time is so deeply ingrained in our experience of the world and in our cultures that we barely think about it apart from the day-to-day grumbles about 'never having enough' or 'how time passes!' Think about the instances in the past day when your thoughts or words have involved time in any sense: wondering if you can stay under the duvet another ten minutes, rushing to catch a train that leaves at 07:50 perhaps, agreeing to meet over coffee 'in five' or suggesting going for a drink next week. This tends, therefore, to fix our attention on only one aspect of time, one means of appreciating it that is constrained and, at times, even lazy: the here and now.

Think back two generations. There are people you know who will have met people born before the American Civil War

Yet with only a little attention we can recalibrate our perspective. To take a good look back in time you need only look up at the sky on a clear dry summer's evening. There you will be greeted with visions of the past: stars as they were tens, hundreds and thousands of years ago: light that, in some cases, left a burning star deep in the

Milky Way before Earth's first civilisations emerged. And you are seeing it now, in the star's future.

Alternatively, think back two generations. There are people you know who will have met people born in the mid-19th century — before the Great Exhibition, the American Civil War, the publication of Darwin's *On the Origin of Species* and long before the invention of aircraft and automobiles. More significantly, flip this forward and you may one day know people (depending, of course, on your current age and life expectancy) who will be alive well into the 22nd century — feasibly in the year 2155 and beyond. The phrase 'future generations' somehow conjures up a world beyond our lifetimes, yet many people live to see their great- and even great-great-grandchildren. How sobering this is when we find it quite difficult to think about the world just five or ten years hence.

Yet there can be no doubt that a longer-term approach to thinking about the world and its sustainability is necessary. Put bluntly, the 'use now, pay later' model of current economic development is using more resources, more quickly than ever before. You are probably aware of the idea that, if the whole planet consumed as Europe does, we would need two or even three planet Earths. Serious and significant change is unavoidable whether we address the problems or not, yet conceiving of such change is as difficult as imagining the world 150 years from now. The human tendency in such situations is often to feel so daunted that, like the proverbial ostrich, we bury our heads in the sand. Yet it will only be through small efforts to imagine time differently, to address the problems at an individual human level, that anything can be done. As the Chinese philosopher Lao Tzu famously stated around 2,500 years ago, 'A journey of a thousand miles starts with a single step.' Small movements can take us a long way and there is no time like the present.

Time, society and the environment

So, how did we get here? One significant change over the last century and a half has been the divorcing of humanity from a local rural environment. In the mid-19th century industrial Britain became the first major country (excluding small city states) to have a population residing predominantly in urban areas. Demographers suggest that now around half the world's population lives in urban areas.

The consequences of mass urbanisation go beyond the significant increases in consumption and travel. Unlike most office-based work, agricultural working patterns are dictated by the seasons — the earliest calendar systems were effectively agricultural technologies — and thus an appreciation of cyclical time patterns and their relation to the local environment is a crucial skill. As the Bible expresses it:

> To every thing there is a season, and a time to every purpose under the heaven:
>
> A time to be born, and a time to die; a time to plant, and a time to pluck up that which is planted.[2]

Meanwhile, as the farmer reads the seasons for the time to plant, the urban worker in a factory or office continues with activities little influenced by whether it is hot and dry or cool and wet outside. While in the rural environment there is an understanding of the seasonality of food, some direct link between the land you live on and the food you eat, in the urban environment this link is broken — summer fruits are available all year round. There is no longer a time for asparagus, a time for strawberries.

The cultural ties binding time, society and the land are disintegrating in the predominantly urban society

Moreover, the cultural ties binding time, society and the land are disintegrating in the predominantly urban society, particularly in a society that has less time for organised religion.

2 Ecclesiastes iii, 1-2.

In an agricultural society the very nature of the year is reinforced by ritual and festive celebrations. Many cultures have winter festivals celebrating the winter solstice. Hanukkah is the Jewish 'festival of lights'; it features the ritualised lighting of candles. Diwali is the Hindu festival of lights. The coming of spring has its rituals (in Korea a three-day ritual is performed by *mansins*, shamans possessed by spirits, to celebrate the first blossoms of spring). Easter links the Christian celebration of resurrection with older celebrations of rebirth and fertility such as those of the pagans — 'the English word Easter and the German Ostern come from a common origin (Eostur, Eastur, Ostara, Ostar), which to the Norsemen meant the season of the rising (growing) sun, the season of new birth'.[3] The ancient Greeks and Romans had harvest deities, as did the Egyptians, Mesopotamians and Celtics, and the Japanese today. If you believe that the fate of the harvest, and thus life, lies with the power of a deity, you celebrate its success with thanksgiving.[4] In the US there is Thanksgiving, when families return 'home' to celebrate (however romanticised) the staple foodstuffs of American-settler lore, the corn and the turkey.

Revd Canon Martyn Percy[5] explains the modern origin of the harvest festival in the UK, dating it back to:

> the Rev Robert Hawker, who, in 1843, building on Saxon and Celtic Christian customs, began to decorate his church at Morwenstow, Cornwall, with homegrown produce. Through the Victorian era, the festival was embellished and romanticised, probably in an effort to counterbalance the influence of the industrial revolution and secularisation.

---

3   F.X. Weiser, *Handbook of Christian Feasts and Customs* (New York: Harcourt, Brace & World, Inc., 1958): 211. Copyright 1952 by Francis X. Weiser; quoted by www.aloha.net/~mikesch/easter.htm.

4   The Greek for thanksgiving gives us the word 'Eucharist' — the Christian celebration of eating bread and drinking wine in remembrance of the gift of life.

5   Revd Canon M. Percy, 'A Harvest of the Spirit', *The Guardian*, 11 September 2004.

He goes on to add that the festival has 'become slowly, but radically, politicised' as 'the emphasis has moved from thankfulness for our abundance to one of concern for those whose experience of provision is one of scarcity, or even outright starvation'.

Still, the fact remains that, for vast swathes of the urban world, harvest festival is a feature from school days and no longer an immediate feature that ties the locality and society to time and the environment. The consequence is seeing these as abstract issues to be addressed by others (politicians, business people, environmentalists) and not as questions to be tackled personally.

So, as the forthcoming chapters will suggest, the environment is a time issue, as are the economy and society. The contributors to this book were asked to consider particular aspects of time, written to stimulate thinking, challenge long-held views and opinions and to provide different perspectives on some age-old questions about the future of the planet. As an edited collection, the authors speak with different voices and from different perspectives, each designed to be read either on its own or as part of the whole; as a consequence, though we've attempted to reduce repetition, in a few places some overlap has been necessary.

Summary of the book

In his opening chapter, the astronomer Martin Rees offers his perspectives on time on the cosmic scale: how can we conceptualise billions of years? If we can do that, where does civilisation fit into the picture? What happens if we can go on for ever? What if we cannot?

He is followed by biologist Ghillean Prance who takes on the idea of long passages of time, explaining how gradual changes in DNA have helped scientists measure the age of species. He goes on to explore how nature has developed genetic clocks intrinsic to the working of many species from Amazonian plants to you and I, and how one species, *Homo sapiens*, seems determined to wreak havoc on the rest.

The problem according to Jonathon Porritt is the world's increasing population, with inevitable consequences for consumption. He calls for politicians to address the issue without recourse to techno-fixes — the assumption that a new technology will solve the problems of an old one. In fact, he contends, the ever greater speed of technology is divorcing humanity from the planet and other people, exacerbating the problems.

Not all humans are divorced from time in nature, however, argues writer Jay Griffiths. She contrasts the modern Western mind-set and its reliance on clocks and Greenwich Mean Time with cultures where time is more flexible and associated with the environment in which people live. As Griffiths writes of indigenous people, their approach to time is 'unpredictable, demanding flexibility, fluidity and quick co-ordination'.

Alexandra Jones and Will Hutton of The Work Foundation take as their starting point the relationship between time and money and posit three ways we could value time differently and more beneficially.

The idea that you can spend time is taken further by New Economics Foundation associate David Boyle. He describes the origins of time banks, arguing that thinking about time as a currency has massive potential implications for more than just volunteering: healthcare, regeneration and the links and connections that make society work as a whole. Sustainability is, of course, about a well-functioning society, not just a functioning environment.

And society as a whole is addressed by Geoff Mulgan, former head of strategy within 10 Downing Street in an updated version of his influential essay on time politics. Time, he suggests, is political: how people access services, run their lives and interact with society.

Public policy does not stand still, however. Last year's thorny ethical question may be widely acceptable some time in the near future. Philosopher Mary Warnock has more experience in this than most, having chaired the Committee of Inquiry into Human Fertilisation and Embryology in 1984. She examines two examples where an awareness of the past and the future (respectively) are critical in developing an approach to an ethical dilemma.

As a number of the contributors note, technology is changing our relationship with time. In the final chapter James Goodman and Britt Jorgensen, with their unique insight into technology and sustainable development, challenge some commonly held perceptions about the impact technology is having on how we handle and experience time. Ultimately, could new technologies make us more sustainable than before?

To round the book off, the Social Market Foundation's Vidhya Alakeson has taken up the challenge of pulling together some of the common themes and recommendations from the book into a conclusion. However, it is time to return to the cosmos . . .

1

Perspectives on time

Martin Rees

About 4.5 billion years have elapsed since our Sun condensed from a cosmic cloud. The proto-Sun was encircled by a swirling disk of gas. Dust in this disk agglomerated into a swarm of orbiting rocks, which then coalesced to form the planets. One of these became our Earth — the 'third rock from the Sun'. The young Earth was buffeted by collisions with other bodies, some almost as large as the planets themselves: one such impact gouged out enough molten rock to make the Moon. Conditions quietened and the Earth cooled, setting the scene for the emergence of the first life.

A memorable early photograph taken from space depicted 'Earthrise', as viewed from a spacecraft orbiting the Moon. Our habitat of land, oceans and clouds was revealed as a thin delicate glaze, its beauty and vulnerability contrasting with the stark and sterile moonscape on which the astronauts left their footprints. We have had these distant images of the entire Earth only for the last four decades. But our planet has existed for more than a hundred million times longer than this.

A brief history of a planet

For more than a billion years, oxygen accumulated in the Earth's atmosphere — a consequence of the first unicellular life. Thereafter, there were slow changes in the vegetation, and in the shape of the land masses as the continents drifted. The ice cover waxed and waned: there may even have been episodes when the entire Earth froze over, appearing white rather than pale blue.

The only abrupt worldwide changes were triggered by major asteroid impacts or volcanic super-eruptions. Occasional incidents like these would have flung so much debris into the stratosphere that for several years, until all the dust and aerosols settled again, the Earth looked dark grey, rather than bluish white, and no sunlight penetrated down to land or ocean. Apart from these brief traumas, nothing happened suddenly: successions of new species emerged, evolved and became extinct on geological time-scales of millions of years.

But in just a tiny sliver of the Earth's history — the last one millionth part, a few thousand years — the patterns of vegetation altered much faster than before. This signalled the start of agri- culture — the imprint on the terrain of a population of humans, empowered by tools. The pace of change accelerated as human populations rose. But then quite different transformations were perceptible; and these were even more abrupt. Within 50 years — little more than one hundredth of a millionth of the Earth's age, the amount of carbon dioxide in the atmosphere, which over most of Earth's history had been slowly falling, suddenly began to rise much faster.

Imagine that a race of scientifically advanced extraterrestrials had been watching our Solar System over the whole 4 billion years. On the previous evidence they might confidently predict that the Earth would face doom in another 6 billion years, when the Sun, in its death throes, will swell up into a 'red giant' and vaporise anything remaining on our planet's surface. But could they have predicted this unprecedented spasm less than halfway through the Earth's life — these human-induced alterations occupying, overall, less than a millionth of the elapsed lifetime and seemingly occurring with runaway speed?

An infinite future?

It will not be humans who witness the Sun's demise 6 billion years hence: it will be creatures as different from us as we are from bacteria. Long before the Sun finally licks Earth's

> It will not be humans who witness the Sun's demise 6 billion years hence: it will be creatures as different from us as we are from bacteria

face clean, post-human intelligence could have spread far beyond its original planet, taking forms that might see the destruction of our Earth as a minor or sentimental matter and still look forward to a longer-range future. The cosmic future extends far beyond the demise of the Sun: the wider cosmos may have an infinite future ahead of it. We can't predict what role life will eventually carve out for itself: it could become extinct; on the other hand, it could achieve such dominance that it can influence the entire cosmos.

Such speculations have generally been left to science fiction writers. But scientists can make some tentative ultra-long-range forecasts. The universe seems fated to continue expanding. Energy reserves are finite, and at first sight this might seem to be a basic restriction. But this constraint is actually not fatal. As the universe expands and cools, lower-energy quanta of energy (or, equivalently, radiation at longer and longer wavelengths) can be used to store or transmit information. Just as an infinite series can have a finite sum (for instance, $1 + \frac{1}{2} + \frac{1}{4} + \ldots = 2$), so there is no limit to the amount of information processing that could be achieved with a finite expenditure of energy. Any conceivable form of life would have to keep ever cooler, think ever slower and hibernate for ever longer periods.

Physicists now suspect that atoms don't live forever. In consequence, long-dead stars and planets will erode away, maybe in a trillion trillion trillion years — the heat generated by particle decay makes each star glow, but as dimly as a domestic heater. Thoughts and memories would only survive beyond this era if downloaded into complicated circuits and magnetic fields in clouds of electrons and positrons — maybe something that would resemble the threatening alien intelligence in *The Black Cloud*, the first and most imaginative of Fred Hoyle's science fiction novels, written in the 1950s.[1]

The endgame could be spun out for a number of years so large that to write it down you'd need as many zeros as there are atoms in all the galaxies we can see. As Woody Allen once said, 'Eternity is very long, especially toward the end.'

We can't predict what role life will eventually carve out for itself: it could become extinct; on the other hand, it could achieve such dominance that it can influence the entire cosmos

Hence, the nature of longevity is change. Darwin himself noted that 'not one living species will transmit its unaltered likeness to a distant futurity'. The great biologist Christian de Duve[2] envisages that:

> The tree of life may reach twice its present height . . . This could happen through further growth of the human twig, but it does not have to. There is plenty of time for other twigs to bud and grow, eventually reaching a level much higher than the one we occupy while the human twig withers . . . What will happen depends to some extent on us, since we now have the power of decisively influencing the future of life and humankind on Earth.

Though time slows, evolution is speeding up. In H.G. Wells's *The Time Machine*[3] the chrononaut gently eased the throttle of his machine forward: 'Night came like the turning out of a light, and in another moment came tomorrow.' As he sped up,

1   F. Hoyle, *The Black Cloud* (Harmondsworth, UK: Penguin, 1957).
2   Christian de Duve, 'Constraints on the Origin and Evolution of Life', in *The Challenges of Sciences: A Tribute to the Memory of Carlos Chagas* (Vatican City: Pontifical Academy of Sciences, 2002).
3   H.G. Wells, *The Time Machine* (London: Heinemann, 1924/1895).

the palpitation of night and day merged into one continuous greyness ... I travelled, stopping ever and again, in great strides of a thousand years or more, drawn on by the mystery of the Earth's fate, watching with a strange fascination the sun grow larger and duller in the westward sky, and the life of the old Earth ebb away.

He encounters an era where the human species has split into two: the effete and infantile Eloi, and the brutish underground Morlocks who operate the factories that make their clothes and goods, but emerge at night to kill and eat them. Finally, he ends up 30 million years hence, in a world where all familiar forms of life have become extinct.

Why the future matters

To a physicist, time is part of the bedrock of reality: a fourth dimension

What happens in far-future aeons may seem blazingly irrelevant to the practicalities of our lives. But the cosmic context is far from irrelevant to the way we perceive our Earth and the fate of humans — indeed it actually strengthens our concerns about what happens here and now, because it offers a vision of just how prodigious life's future potential could be.

In our everyday life, time is a commodity. We gain or lose it; we save or spend it; all too often we merely waste it. But to a physicist, time is part of the bedrock of reality: a fourth dimension. We are used to the three dimensions of space. Three numbers are needed to define a location on Earth: latitude, longitude and elevation. But to specify a 'happening' we need a fourth number as well — the number that tells us when the event happens. There is, however, a crucial difference between time and the three spatial dimensions. We can move to left or right, forwards or backwards, up or down. But we are carried relentlessly forward in time. Time machines that allow us to revisit the past are the stuff of fantasy.

My professional interest is in the science of the entire cosmos — I study our environment in the widest conceivable perspective. This might seem an incongruous viewpoint from which to focus on practical terrestrial issues. But a preoccupation with near-infinite spaces doesn't make cosmologists especially 'philosophical' in coping with everyday life; nor are they less engaged with the issues confronting us here on the ground, today and tomorrow. My subjective attitude was better expressed by the mathematician and philosopher Frank Ramsey:[4]

> I don't feel the least humble before the vastness of the heavens. The stars may be large, but they cannot think or love; and these are qualities which impress me far more than size does . . . My picture of the world is drawn in perspective, and not like a model drawn to scale. The foreground is occupied by human beings, and the stars are all as small as threepenny bits.

Our normal 'time horizons' are of course very limited. Economic decisions generally discount into insignificance what may happen more than 20 years from now: commercial ventures aren't worthwhile unless they pay off far sooner than that, especially when obsolescence is rapid. Government decisions are often as short-term as the next election. But sometimes — in energy policy, for example — the horizon extends to 50 years.

The debates about global warming that led to the Kyoto Protocol take cognisance of what might happen 100-200 years ahead: the consensus is that governments should take pre-emptive actions now, in the putative interest of our 22nd-century descendants (though whether these actions will actually be implemented is still unclear).

There is just one context where public policy looks even further ahead, not just for hundreds but for thousands of years: the disposal of radioactive waste from nuclear power stations. Some of this waste will remain toxic for thousands of years, and, both in the UK and the US, the specification for underground depositories

4   F. Ramsey, paper given to the Apostles' Society, 1925.

demands that hazardous materials should remain sealed off — with no leakage via groundwater, or through fissures opened up by earthquakes — for at least 10,000 years.

The prolonged debates on radioactive waste disposal have had at least one benefit: they have generated interest and concern about how our present-day actions resonate through several millennia. These time spans extend far beyond the horizon of most other planners — but they are still infinitesimal compared to the future lifetime of the Earth itself.

Human determinism

Most people nonetheless have little conception or awareness of the far future: humans are often tacitly regarded as the culmination of evolution. Traditional Western culture envisaged a beginning and an end of history, with a constricted time span in between. For centuries there was broad acceptance that Earth had existed for a few thousand years, and that humans had appeared on the scene soon afterwards. Moreover, history was widely believed to have already entered its final millennium. For the 17th-century essayist Sir Thomas Browne, 'the world itself seems in the wane. A greater part of Time is spun than is to come.' However, we know differently now.

Technical and environmental change has been accelerating over human history. A Neanderthal woman would have expected her children to live out their days in a similar way to her own generation, as indeed would most humans until medieval times. Substantial progressive change within a single lifetime is a distinctive hallmark of recent centuries: some technologies now advance so fast that all hardware is scrapped (or consigned to museums) within a very few years. Biotechnology is now, quite suddenly, opening up an unprecedented new dimension of change: the biological fundamentals of humanity, essentially unaltered throughout recorded history, could be transformed within a century.

In Wells's story, it takes 800,000 years for humans to divide into two subspecies — a time span that accords with modern ideas of how long it took for humanity to emerge via natural selection. But in the new century, changes in human bodies and brains won't be restricted to the pace of Darwinian selection, nor even to that of selective breeding. Genetic engineering and biotechnology, if widely practised, could transmogrify humanity's physique and mentality far faster than Wells foresaw. Our own species may change and diversify faster than any predecessor — via intelligently controlled modifications, not by natural selection alone. Already we are able to screen for genetic illnesses and select embryos that are free of specific gene mutations.

In the new century, changes in human bodies and brains won't be restricted to the pace of Darwinian selection

Life beyond Earth

In future centuries, robots and fabricators could pervade the entire Solar System. Whether humans will themselves have joined this diaspora is harder to predict. If they did, communities would develop in a manner that eventually made them quite independent of Earth. Unconstrained by any restrictions, some would surely exploit the full range of genetic techniques and diverge into new species. The diverse physical conditions — very different on Mars, in the asteroid belt and in the still colder far reaches of the Solar System — would give renewed impetus to biological diversification.

The entire Galaxy, extending for 100,000 light years, could be 'greened' in less time than it took for us to evolve from the first primates. Far-future life could display even more variety than has been played out in the entire annals of the Earth's biosphere. Artefacts created by us, and in some way descended from us, may use their own intelligence to develop further, on Earth and far beyond.

A key challenge for science is to understand the nature of life — how it began, and whether it exists beyond Earth. (There is certainly

Far-future life could display even more variety than has been played out in the entire annals of the Earth's biosphere no other scientific question which I personally would be more eager to see answered.) Alien life may be discovered — even, conceivably, alien intelligence. Our planet could be one of millions that are inhabited — we may live in a biofriendly universe already teeming with life. If so, the most epochal happenings on Earth, even our utter extinction, would barely register as a cosmic event. In the words of the 18th-century astronomer and mystic, Thomas Wright of Durham:[5]

> In this great Celestial Creation, the Catastrophy of a World, such as ours, or even the total Dissolution of a System of Worlds, may possibly be no more to the great Author of Nature, than the most common Accident in Life with us, and in all Probability such final and general DoomsDays may be as frequent there, as even Birth-Days or Mortality with us upon this Earth.

But we might learn enough to conclude that intelligent life is rare, or even that the infinite cosmic spaces are still sterile voids — this is certainly a tenable viewpoint at the moment. Earth's biosphere could be (quite plausibly) the unique abode of intelligent and self-aware life within our Galaxy. Were that so, our small Earth's fate would have a significance that was truly cosmic — an importance that would reverberate through the whole of Thomas Wright's 'celestial creation'.

The first aquatic creatures crawled onto dry land, in the Silurian era more than 300 million years ago. They may have been unprepossessing brutes but, had they been snuffed out, the potential of land-based fauna would have been jeopardised. Likewise, the post-human potential is so immense that not even the most misanthropic among us would countenance its being prevented by human actions.

Our concerns naturally focus more insistently on the fate of our present generation than on prospects for the far future. But for me, and perhaps for others (especially those without religious belief),

---

5   T. Wright, *An Original Theory of the Universe* (London, 1734).

The Clock of the Long Now

It has been called 'the Millennium Clock', 'the World's Slowest Computer' and 'a gigantic piece of retro technology';[6] musician Brian Eno named it 'The Clock of the Long Now': in the Nevada Desert, a visionary foundation is building a clock to last 10,000 years. The half-scale prototype currently residing in London's Science Museum stands one-and-a-half feet taller than Big Ben. The final model will tick once a year, chime once a century and be accurate to within a day in 20,000 years.

Inventor and engineer Danny Hillis first conceived of this monumental project in 1993. 'The idea of the Clock is to encourage long-term thinking, which is in short supply these days,' says scientist Stewart Brand, erstwhile eco-warrior and president of the Long Now Foundation. The Foundation aims to counter what it sees as 'today's "faster/cheaper" mind-set'. The Clock of the Long Now evokes not so much a digital watch but the pyramids or Stonehenge, which had both a practical and a spiritual function. It is a 'living monument'[7] for an age that values universal human rights above religious institutions.

Although Hillis once designed new theme park rides for Disney, the clock is no gimmick. It uses only mechanical, not digital, principles and its cutting-edge design incorporates tungsten carbide and synthetic sapphire. It 'utilises a new form of digital calculation and synchronises with the noon sun', harking back to some of our very earliest timekeeping devices, obelisks built by the Egyptians in 3,500 BC which functioned as primitive sundials.

Jonas Salk commented to Hillis, 'You want to preserve something of yourself [by building this]', but the clock will do more than preserve one man's work. It will embody some of the principles of sustainability. Its design embraces change: Hillis allows for such unpredictable factors as human curiosity, theft, bad weather and earthquakes. It should be possible 'to improve the clock with time', and 'to determine its operational principles' through observation — archae-

6 'Whole Earth Visionary', Guardian Unlimited, www.guardian.co.uk/saturday_review/story/0,,531898,00.html.
7 S. Brand, The Clock of the Long Now: Responsibility and Time (New York: Basic Books, 1999).

ologists would have an easier time if Stonehenge or the Nazca Lines in Peru were as simple to understand, thousands of years after their creation.

The Long Now Foundation runs a number of projects with wide-ranging implications for sustainable development. The Rosetta Disk would function as a 'long-term linguistic archive and translation engine, designed to allow for the "recovery" of lost languages in the deep future'; highly relevant when 'Linguists estimate that . . . half of the world's languages will disappear in the next 100 years,' according to Professor Peter Austin of the School of Oriental and African Studies. The Long Bets website allows people to make 'societally or scientifically important' predictions about events to take place at least two years in the future for a modest $50 charitable donation a head — a guilt-free way to have a flutter. All these projects encourage us to think about our lives in the context of thousands of years of human history, some of which hasn't happened yet.

In 1993, Hillis wrote: 'Maybe the clock is my way of explaining [that] I cannot imagine the future, but I care about it . . . I sense that I am alive at a time of important change, and I feel a responsibility to make sure that the change comes out well . . . I have hope for the future.'[8] The Clock of the Long Now is one small step for Danny Hillis, but one big step towards bringing the future into the present, and ensuring that what we build, we build to last.

Mireille Kaiser

---

8   D. Hillis, 'The Millennium Clock', *Wired* 3.11 (1995); available online at www. wired.com/wired/scenarios/clock.html.

this longer-term perspective strengthens the imperative to cherish this 'pale blue dot' in the cosmos, and not foreclose life's long-range future.

Our challenge

Think back to the hypothetical aliens: if they continued to keep watch on the history of the Earth, what might they witness in the next 100 years? Will a final squeal be followed by silence? Or will the planet itself stabilise? And will some of the small metallic objects launched from the Earth spawn new oases of life elsewhere in the Solar System, eventually extending their influences — via exotic life, machines or sophisticated signals — far beyond the Solar System, creating an expanding 'green sphere' that eventually pervades the entire Galaxy?

Many of us are less confident that our civilisation will survive the next century than were our forebears, who devotedly added bricks to cathedrals that would not be finished in their lifetime

The evolution of Earth's biosphere can now be traced back several billion years: the future of our physical universe is reckoned to be more extended still, perhaps even infinite. But despite these expanded horizons, both past and future, one time-scale has contracted: many of us are less confident that our civilisation will survive the next century than were our forebears, who devotedly added bricks to cathedrals that would not be finished in their lifetime. What happens here on Earth, in this century, could conceivably make the difference between a near-eternity filled with ever more complex and subtle forms of life, and one filled with nothing but the elements.

2

Natural clocks

Ghillean Prance

As dusk arrives by the bank of an Amazonian lake, the white flowers of the royal water lily open like stars across the water. Immediately, as they begin to open, scarab beetles start to arrive and enter the flowers to feed. The flowers are primed to open at night just as their nocturnal pollinator takes to the air. Once the beetles are inside, the flowers close up and trap the insects within. During the next day the flowers change colour from white to red, in evening they reopen and precisely as they do the pollen is released. The beetles emerge from the base of the central cavity of the flower all sticky from plant juices and the pollen adheres to them. They fly off to find another white flower, and since the water lily produces flowers every second day they will carry the pollen to a different plant and thereby cross-pollinate it — a perfect example of how the timing and co-ordination of events is intrinsic to nature.

Nature is full of such interactions between organisms that require precise timing to be effective. These are adaptations to the natural cycles of seasons and night and day and they have slowly evolved over countless generations. Today, however, things are changing at an unprecedented pace through the intervention of a single species, *Homo sapiens*. The rate of change is too rapid for plants and animals to adapt to and this is one of the many threats to the diversity of living organisms on our planet.

In its short time on Earth *Homo sapiens* has rapidly begun to intervene with biological time. This latecomer species has had more adverse effect than any other and is now precipitating a massive wave of species extinction comparable to that which wiped out the dinosaurs and many other organisms about 65 million years ago. Humans are altering natural habitats and converting them into fields and cities; they are causing severe disruption to nature through pollution and by causing the world's climate to change. Biology cannot keep up with this because natural changes have usually occurred at a much slower pace, thus allowing for the process of evolutionary adaptation to take place.

Homo sapiens is now precipitating a massive wave of species extinction comparable to that which wiped out the dinosaurs

Long-term biological time

We now are able to estimate with reasonable accuracy the pace of evolution over the geological ages by studying the evolution of the DNA molecule. Genes are made from two chains of proteins that create the famous double helix: one containing cytosine and thymine, and the other adenine and guanine. It is from these four proteins that we get the CTAG patterns used to describe the coding of genes. The order of the code informs the creation of different proteins required by living organisms, acting like cogs and switches, interacting with one another to produce different effects.

Imagine two different species: the longer they have been evolving separately, the more amino acid differences accumulate in their proteins and this reflects mutation of the genes or segments of DNA that control their formation. Although the basic rate of mutation is probably quite similar for all genes, the process of natural selection removes those mutations that impair the function of a protein. These functional constraints affect the rate at which amino acids change in any particular protein. Different proteins change at very different rates over the more than a billion years of evolutionary time.

For example, the structure of the protein histone is so rigidly defined for its DNA-binding function that in the one billion years since plants and animals separated, only one amino acid difference exists between pea histone and cow histone. On the other hand, there are fibrinopeptides — segments of the fibrinogen molecule consisting of about 20 amino acids. These are important in blood clotting and change rapidly: there is an 86% difference in the amino acids of fibrinopeptides in the horse compared with the human. Each protein has a characteristic rate of change and this can be used together with fossil evidence to pinpoint the timing of events within the evolutionary timetable. Histones undergo change once in a billion years whereas fibrinopeptides average one mutation every million years: one thousand times more quickly.

Molecular clocks run at slightly different rates in different organisms and so generalisations cannot be made. The same protein may change faster in one group of organisms than in another. In order to estimate the most accurate time of divergence between species, where possible, biologists check the rate of molecular change against the fossil record. Molecular clocks are not precise timepieces, but, as more and more molecules are compared, it is possible now to get a good idea of the timing of many evolutionary changes. This means that in addition to the fossil evidence we now have a molecular chronology of evolution and can estimate the pace of changes and adaptations of many organisms. This shows a long, slow pathway from simple unicellular life to the complex biodiversity we have today.

Jet lag and plant switches

It is not merely in the molecular structure of genes that changes suggest natural clocks; at much shorter time spans, plants are highly responsive to the daily cycle of light and darkness. Some years ago, I carried out some experiments on the royal water lily, which showed that red light makes the flower open more rapidly than ordinary white light does. On the other hand, far-red light (created by enclosing the flowers in red and blue cellophane bags for two hours before nightfall) causes inhibition of flower opening. This showed us that a particular chemical — phytochrome — acts like a switch, responding to the light at a particular time of day, telling the flower when to open and close. Phytochrome occurs in every major group of plants and is linked to environmental light signals, controlling many activities of plants such as flowering, germination and the circadian rhythms.

This sensitivity to the 24-hour daily cycle is not confined to plants. I write this just after a journey from a part of North America that was six time zones away from my native UK. Anyone who travels knows the feeling of jet lag. This occurs because the human body is also set to a 24-hour clock, known by scientists as circadian rhythm. This term originated from the Latin *circa dies*, meaning 'about one day'. An internal biological clock is fundamental to all living organisms. In humans it influences hormones that play a role in sleep and wakefulness, metabolic rate and body temperature. Disruption of circadian rhythms not only affects sleep patterns, but has also been found to precipitate mania in people with manic-depressive illness. Other types of illnesses also are affected by circadian rhythms. For example, heart attacks occur more frequently in the morning while asthma attacks occur more often at night. The level of the hormone melatonin rises in our bodies during the night and falls during the day because of the circadian rhythm.

The circadian rhythms are genetically controlled. The first circadian gene was discovered in the fruit fly in 1971 and a second 13 years later. In 1997 a circadian gene was first found in a mammal, the

An internal biological clock is fundamental to all living organisms mouse. This discovery accelerated the search for other clock genes, and findings in higher-order mammals are yielding a consistent picture of the role and function of circadian rhythms in organisms from bacteria to plants to animals. We know most about the process in fruit flies.

The interaction between four regulatory proteins stimulated by light creates the daily rhythm of the fruit fly's clock. These proteins are made by genes in the DNA and act like switches. Early in the day, two genes called 'cycle' and 'clock' produce two proteins called CYCLE and CLOCK.[1] When these two proteins bind to one another, they turn on the genes 'per' (short for period) and 'tim' (short for time) that make PER and TIM proteins. These PER and TIM proteins accumulate over several hours until late in the day when they reach levels that turn off CYCLE and CLOCK. This in turn slows down the production of PER and TIM, which begins the cycle all over again on a daily basis (see Figure 1). Recent research on circadian rhythms has

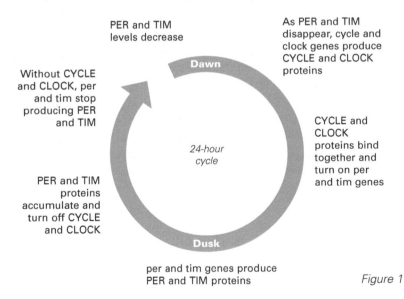

PER and TIM
levels decrease

As PER and TIM
disappear, cycle and
clock genes produce
CYCLE and CLOCK
proteins

Without CYCLE
and CLOCK, per
and tim stop
producing PER
and TIM

CYCLE and
CLOCK
proteins bind
together and
turn on per
and tim genes

PER and TIM
proteins
accumulate and
turn off CYCLE
and CLOCK

Dawn

24-hour cycle

Dusk

per and tim genes produce
PER and TIM proteins

Figure 1

1 Geneticists give the gene the same name as the protein it creates but differentiate them by writing the gene in lower case and the protein in upper case.

demonstrated a remarkable similarity in mechanisms across such diverse groups of organisms as fungi, plants, insects and mammals. Variations have been found in the human clock gene between different individuals, which may predispose people to be 'early birds' or 'night owls'.

The pace of modern life, jetting around the world and regularly taking part in activities that interrupt our pattern of sleep, goes against our genetic heritage which, through the process of evolutionary adaptation, has programmed us and other organisms to the 24-hour daily cycle. We must beware of the impact of interrupting this natural timing cycle. Apart from the effect of our acceleration of time on the natural ecosystem, we are also damaging ourselves by not adhering to natural circadian rhythms.

> Variations have been found in the human clock gene between different individuals, which may predispose people to be 'early birds' or 'night owls'

Time for climate change

Changes in climate are natural phenomena that are well documented, such as the various ice ages that several times covered northern Europe with blankets of ice and snow. However, over the last four decades the climate has begun to change at a much more rapid rate. According to the IPCC (Intergovernmental Panel on Climate Change) by 2000 global average temperatures had risen by about 0.4°C above the 1961–90 average and by about 0.8°C since the beginning of the 20th century.[2] There is little doubt that this is because of human activities. It is now evident that the increase in carbon dioxide and other greenhouse gases such as methane is causing this change. According to the US National Oceanic and Atmospheric Administration, levels of carbon dioxide in the atmosphere have risen from about 315 ppm (parts per million) in the mid-

2 *New Scientist*, 22 March 2004; www.newscientist.com/hottopics/climate/climate.jsp?id=ns99994802.

1950s to 375 ppm today.[3] The IPCC estimates that, by 2100, levels may reach between 650 ppm and 970 ppm. This is caused mainly by the burning of fossil fuels and by deforestation, particularly of tropical rainforest. Not only is the average worldwide temperature increasing, violent storms are occurring more frequently and sea levels are slowly rising due to melting of glaciers and thermal expansion of the oceans. For the natural world one of the main problems of climate change is the rate at which it is occurring. Plants and animals are not able to keep pace with this rate of change.

Recent scientific literature contains many papers about the effects of climate change on natural and artificial ecosystems (for some examples, see page 35). A study carried out over the last 50 years at the Royal Botanic Gardens, Kew, by botanist Nigel Hepper,[4] shows that on average plants are now blooming six days earlier than when he began his observations. Another study reports that plants around Washington, DC, flowered six days earlier in 1993 than in 1959.[5] In western Canada the aspen is blooming 26 days earlier than a century ago. At the same time migratory birds are arriving in the north considerably earlier than 50 years ago and then are leaving later in the autumn. In the UK both the National Trust and the Royal Horticultural Society announced recently that they are having to rethink their planting schedules to use plants adapted to warmer climates. It might seem good to be able to grow more exotics in our gardens, but there are various downsides to this. Some new introductions are likely to escape from gardens and become invasive species, as the

Migratory birds are arriving in the north considerably earlier than 50 years ago and then are leaving later in the autumn

---

3   F.N. Hepper, 'Phenological Records of English Garden Plants in Leeds (Yorkshire) and Richmond (Surrey) from 1946 to 2002: An Analysis Relating to Global Warming', *Biodiversity and Conservation* 12 (2003): 2,503-20.

4   M.S. Abu-Ashu, P.M. Peterson, S.G. Shetler and S.S. Orli, 'Earlier Plant Flowering in Spring as a Response to Global Warming in the Washington, DC, area', *Biodiversity and Conservation* 10 (2001): 597-612.

5   G.-R. Walther, E. Post, A. Menzel *et al.*, 'Ecological Responses to Recent Climate Change', *Nature* 416 (2002): 389-95.

Japanese knotweed and the waterweed *Crassula* have already done. These have devastated local biodiversity, creating monocultures where only they survive. In the UK it has been illegal to spread Japanese knotweed since the mid-1980s.

The milder winters also mean that many insect pests will survive throughout the year. Alien insects are finding it easier to survive and so there is a real danger of the vectors of diseases such as malaria returning to northern Europe. Already there is evidence that a steady rise in annual temperatures is associated with expanding mosquito-borne diseases in the highlands of Asia, East Africa and Latin America.

The danger of diseases not only affects humans but also our crops, as new crop pathogens arrive on our shores. Phenology is the study of the effect of climate on the seasonal events of flora and fauna such as dates of flowering, leaf flush in plants and migratory events in animals. Changes in phenology are serving as natural indicators that climate change is occurring.

Some of the plants most threatened by climate change in the UK are our alpines — plants living above the level of trees at higher altitudes. As the climate warms, the zone of alpine climate gradually moves higher up the mountains until eventually it no longer exists. The special climate to which these sensitive plants are adapted is disappearing and they are likely to become extinct in the UK. Reports from Alaska show that there is an expansion upwards of shrubs in previously shrub-free areas. Plants are less mobile than animals and they may not be able to migrate rapidly enough to keep up with the rate of climate change. This is further exacerbated by the way in which so many natural ecosystems have been broken up into fragments rather than existing as continuous vegetation. This becomes a barrier for the dispersal of many plants and indeed animals too. All relatively slow-breeding species of plants and animals will have trouble adapting to climate change. The rate is so much faster than natural change that there is not enough time for mutations and subsequent adaptation to the new circumstances.

Small natural and gradual changes in climate will neither alter dramatically the composition of an ecosystem nor disorder its functions. However, the effect of rapid human-induced changes of the seasonal cycle can be significant because they disrupt the way in which whole ecosystems respond and interact.

Changes in biological timing may be more serious still. Winter warming has precipitated breeding season changes in some British species of amphibians. This variability has, in turn, altered what biologists term 'temporal niche overlaps' in breeding ponds, with adverse consequences for how different species feed. The timing of arrival of migrants or of the breeding of amphibians is often separated in ways that ensure that the necessary food supplies are available. With birds arriving earlier in spring, the competition for nest sites may increase as well. The warmer weather in spring in Europe has disrupted the synchrony between winter moth (*Operophtera brumata*) hatching and oak bud burst, leading to a mismatch between the time in the day when the insect is available and the time in the day when its predator, great tit nestlings, demand food. Timing of events is fine-tuned and vital in nature and this has evolved over many generations in order to divide and use the available resources to their maximum. Human-induced changes are interrupting nature's chronology and causing many aspects of its intricate interactions to collapse.

Human-induced changes are interrupting nature's chronology and causing many aspects of its intricate interactions to collapse

Time for action

We think of time in seconds, minutes and hours. However, living organisms have adapted to the daily cycle of the earth's rotation around its axis and to the seasonal cycle resulting from the orbit of the Earth around the Sun. All the changes in nature that are happening around us today should be a strong warning that all is not well with the new time-scale that is being forced on it. The major effects of a relatively small climate change — on alpines, on winter moths, on sea levels and average temperatures — have mostly happened in the last three decades: a very short time span indeed. This should be a warning that we have little time to correct the situation if we wish to avoid major extinction. If we want to create a sustainable future then climate change must be addressed, for the biological clocks and cycles and the pace at which evolution proceeds cannot be easily altered. We ignore this warning at our peril because we humans too depend on our interactions with the rest of nature for our survival.

Titles of recent scientific articles indicating unusual phenological events affecting the biology of organisms in many places

- 'Environmental Change and Antarctic Seabird Populations', *Science* 297 (August 2002): 1,510.

- 'Reorganisation of North Atlantic Marine Copepod Biodiversity and Climate', *Science* 296 (May 2002): 1,692.

- 'Climate Warming and Disease Risks for Terrestrial and Marine Biota', *Science* 296 (June 2002): 2,158.

- 'Future Projections for Mexican Faunas under Global Climate Change Scenarios', *Nature* 416 (April 2002): 626.

- 'Increasing Shrub Abundance in the Arctic', *Nature* 411 (May 2001): 546.

- 'Climate Change Transforms Island Ecosystem', *Nature* 410 (April 2001): 616.

- 'Climate Impact of Tropical Lowland Deforestation on nearby Montane Cloud Forests', *Science* 294 (October 2001): 584.

- 'Field Studies on the Effects of Global Warming on Mountain Vegetation in Japan', *Global Environmental Research* 1 (1998): 71.

- 'Global Warming Could Mean That More Tender Plants Can Be Grown in the UK', *Horticulture Week*, 3 February 2000.

Time/Age/Speed:
from the kneejerk reflex to the age of the Earth

>> 50 milliseconds

Sit in a comfortable position, with your legs crossed. Now tap just below the knee. Your leg kicks out automatically: you have just demonstrated the kneejerk reflex. Called monosynaptic because it only involves one synapse, the reflex occurs in 50 milliseconds, before the brain has consciously registered any sensation of touch.

>> 1 quarter of a second

It takes a quarter of a second to blink, something we normally do 20 times per minute. We blink more frequently when nervous or excited; less often when working in front of a computer. In total, we blink over 10,000,000 times a year. The blink reflex performs an essential function: it keeps the surface of the eye clean and moist. The 'lost time' of a blink keeps our sight clear and our vision sharp.

>> 1 hour

In one hour, some varieties of bamboo can add as much as 5 cm to their height. Bamboo, a type of grass and one of the fastest-growing plants, is also very hardy, with a tensile strength superior to that of mild steel. In 1945, a bamboo plant was the living thing nearest to ground zero in Hiroshima that survived the blast. It sent up new shoots soon afterwards.

>> 1–2 days

The oldest winged insect is the proverbially short-lived mayfly (of the order of Ephemeroptera, whose name comes from the Greek *ephemeros*, 'lasting a day'). Pausing neither to eat nor to court, mayflies emerge from the nymph stage with all the food they will need for their adult life, and mate in flight. Typically they live only 1 or 2 days, but the life expectancy of some species can be as long as 2 weeks, or as little as 2 hours. The shortest biography is said to be that of the mayfly: 'Born. Sex. Die.'

>> 13 years

When you first see a child with progeria it seems like a cruel trick of nature; for children with this condition appear to age prematurely, without ever having gone through youth. Progeria is a rare but fatal genetic disease. It occurs in around 1 in 4–8 million births, and, like ageing itself, affects all ethnicities, male and female alike. Children with progeria suffer from growth failure, hair loss, aged skin, osteoporosis and arthritis. They die at an average age of 13, from complications of arteriosclerosis or cardiovascular disease. There is no known cure, but scientists have recently identified the gene responsible for the condition, and hope that this may help to develop a treatment, and provide insights into the natural ageing process.

>> 188 years

There may be some truth in La Fontaine's fable of *The Tortoise and the Hare*. Tortoises, who can survive for long periods without food or water, move extremely slowly, but have been known to live exceptionally long lives. In 1773 or 1777, Captain Cook presented the Tongan royal family with a Madagascar radiated tortoise named Tui Malila. The animal was photographed with Queen Elizabeth and the Duke of Edinburgh in 1953 and died in 1965, aged at least 188 years. Timothy the tortoise, who lived to the age of 160, was a ship's mascot in the Crimean War and until his death in 2004 was thought to be the UK's oldest resident. In the words of Confucius, 'It does not matter how slowly you go so long as you do not stop.'

>> 4,800 years

High up California's White Mountains, there is little water, warmth or sunlight. The soil is poor in nutrients; it is an inhospitable environment. Yet at an altitude of 10,000 feet, a forest of bristlecone pine houses some of the oldest trees on earth. One of them, nicknamed the 'Methuselah Tree' (after the biblical character said to have lived 969 years), discovered by scientist Edmund Schulman, was found to be almost 48 centuries old. The pines have survived so long thanks to an exceptional ability to survive in adverse conditions: in times of drought, they 'shut up shop' and, like Sleeping Beauty, 'reawaken . . . in a favorable year', wrote Schulman.[6] Growing slowly but persis-

6 *National Geographic*, March 1958.

tently, the Methuselah tree has outlived many civilisations and, if it survives the current human threat, may yet outlive many more.

>> 300 million years

Diamonds are the hardest known naturally occurring material, and under normal circumstances do not decay, although, since they can burn or split, they may not quite last forever. Diamonds are carbon crystals that form in unique conditions, under extreme pressures and high temperatures, hundreds of kilometres below the Earth's crust. It takes a few months to create a diamond artificially, but in nature it takes millions of years for a diamond to form. Research suggests that one diamond found in Siberia may have grown over a period of more than 300 million years. The age of the stone, however, is far greater: many diamonds we find at the surface of the earth or in mines are over 3 billion years old.

>> 4.5 billion years

A literal reading of the Bible suggests that the age of the Earth is about 6,000 years, but geological dating tells us that it is around 4.5 billion years old. Plate tectonics have destroyed the oldest rocks on Earth, but ancient rocks more than 3.5 billion years old can be found on all the continents. The oldest dated Moon rocks are 4.4–4.5 billion years old. This pales in comparison with the universe, however, which evidence from the WMAP (Wilkinson microwave anisotropy probe) satellite suggests it is 13.7 billion years old.

Mireille Kaiser

3

Too many, too fast?

Jonathon Porritt

It's the speed with which human numbers have grown over the last 250 years that lies absolutely at the heart of today's ecological crisis.

It is of course the height of political incorrectness to press such an assertion, but in 30 years of campaigning for environmental and social justice issues I've yet to meet a single person who has been able to explain to me how the facts of population growth allow for any other interpretation. People reading a book such as this will 'sort of' know the facts about population. Just as an *aide-mémoire*, four milestones are all we need to grasp the overall problem. In 1750, at the dawn of the Industrial Revolution in Western Europe, total human numbers across the globe amounted to around 500,000,000 — half a billion people.

> There is now almost universal consensus that there will be around 9 billion people on Earth in 2050

Only 200 years later, by 1950, that had grown to 3 billion. It took just 50 years for that figure to double to 6 billion. Track out to 2050, and there is now almost universal consensus among demographers that there will be around 9 billion people on Earth — give or take a few tens of millions.

That slowing-down over the next 50 years (which is of course very different from anything resembling a net decline in itself!) is accounted for by the welcome reduction in average annual growth from its peak of 2.08% in 1965 down to a rather more manageable 1.14% in 2003. But it's the distribution of that next 3 billion that adds an extra — and still largely ignored — dimension. Of the 70 million or so extra people arriving among us every year, the vast majority will be born in developing countries.

From a planetary perspective, however, the real timing crunch remains the same: from 3 billion human beings to 9 billion human beings in just 100 years

While there will be some growth in a few developed countries (and particularly in the United States, which could well see its population up to around 400 million by 2050), the predominant pattern in Europe and Japan will be one of declining numbers. The major growth will be in countries such as India, Bangladesh, Indonesia, Nigeria and most sub-Saharan countries. Total numbers in the developing world will grow to around 8 billion, in contrast to little over 1 billion in the developed world.

From a planetary perspective, however, the real timing crunch remains the same: from 3 billion human beings to 9 billion human beings in just 100 years.

It's mind-boggling that the combined intellectual firepower of the global environment movement, the human rights movement, the development movement and of practically every other progressive cause around the world continues to ignore the adamantine reality of that massive population surge in what amounts to no more than the merest sliver of time since life on Earth first began around 4 billion years ago.

In 1998, the Foundation for Global Community produced the most compelling visual representation of what 100 years means in that evolutionary context: *A Walk through Time: From Stardust to Us*[1] compacts those 4 billion years into a walk covering precisely one mile. Each foot represents around 1 million years. On such a scale,

---

1   S. Liebes, E. Sahtouris and B. Swimme, *A Walk through Time: From Stardust to Us* (New York/Chichester, UK: John Wiley, 1998).

the last Ice Age ends precisely 0.1 of an inch before the end of the walk, and the last 100 years (with its 6 billion people extra people) equates to no more than 0.001 of an inch in that mile.

We're not good at this kind of time-scale. For most of us, the mind just shuts down in the presence of such immensity. We can just about cope with the idea of the Earth being populated by dinosaurs — a mere 65 million years ago — but anything before that is so alien a temporal zone as to deter access to all but a tiny academic minority. And, whatever sense of futurity we may have, it's almost exclusively anthropocentric, inasmuch as we struggle to conceptualise the very idea of life on Earth without humankind being at the heart of it. Rationally, however, we know full well that the prospect of some evolved variant of humankind still being around in another 2–3 billion years (or however long the Sun's energy sustains this particular solar system) is as close to zero as we are capable of imagining.

In that somewhat terminal respect, pedants may well argue that literally nothing is sustainable in that *all* life in this solar system will at that point be extinguished. Unless, of course, you're one of those strange people who find solace in science fiction scenarios that see humankind perpetuating itself not just off-planet but across the universe. To me, that rather seems to be missing the point. As writer Brian Swimme[2] says:

> The origin of the universe is an immense fire that will never again appear in the universe's expansion into time. This primal universe shines forth with such titanic radiant energies that we humans evolving out of this blaze can still feel its heat 15 billion years later. We have been conditioned to regard time in its mechanical and watchlike aspects, almost independent of the universe itself; but a journey through evolutionary time reveals something more. Time is not primarily a clock, nor is time simply an abstract measurement. A meditation on the billions of years of the universe's process provides a glimpse into time as a measure of the universe's creativity.

2 Liebes *et al.*, *op. cit.*.

We are all, in varying degrees, in thrall to the absolutism of the now, and the lion's share of our economic, scientific and political energy is dedicated to the shortest of short-term 'fixes'

As other authors in this collection have commented, it's sometimes hard to detect a great deal of enthusiasm for that extraordinary vista of the universe's creative energy unfolding over a 15-billion-year time span! We are all, in varying degrees, in thrall to the absolutism of the now, and the lion's share of our economic, scientific and political energy is dedicated to the shortest of short-term 'fixes' for sorting out what is perceived to be going wrong in the world today.

The danger of technological fixes

More often than not, these short-term fixes, be they technological or political, are upon us before we've had time to accommodate to the sequence of fixes that came before! Nuclear power is but one example of the problem. It may indeed be something of a cliché, but technology really is speeding up. For me personally, being a member of the Working Group set up by the Royal Society and the Royal Academy of Engineering to carry out an independent study into current and future developments in nanotechnology was an absolute eye-opener. Although it's true that chemists have been creating new polymers (which are basically large molecules made up of very small sub-units) over many decades, and that technologists have been fashioning nano-scale features on computer chips for more than 20 years, there's been a step change over the last few years in the tools that allow atoms and molecules to be examined and manipulated at the nano-scale. (By which, incidentally, we're talking of stuff going on at one thousand millionths of a metre — a single human hair is about 80,000 nanometres wide.)

Though our report in July 2004 definitively set to one side sensationalist fears that the world was about to be reduced to a mass of 'grey goo', with trillions of self-replicating nano-robots (as in

Michael Crichton's page-turner, *Prey*[3]) laying all to waste, it emphatically exhorted government to do three things about the imminent onrush of new applications and processes: first, take time out to fill the yawning knowledge gaps, as we really know very little about the potential impact of nanoparticles on both human health and the environment. Second, fill the regulatory gaps that were revealed during the course of our enquiry, as these are an inevitable consequence of the speed with which new technological innovation impacts on old systems. And, lastly, adopt an unapologetically precautionary approach.

That combination of research, regulation and precaution is all this generation has at its disposal as it seeks both to avoid unanticipated adverse consequences from new technologies for itself, and to do the right thing by way of its responsibilities to future generations. As far as sustainable development is concerned, it's that notion of intergenerational equity that marks it out from all those other notionally 'big ideas' that jostle for attention in today's febrile political marketplace. Though I have to admit that I find the standard definition of sustainable development in the *Brundtland Report*[4] hopelessly defective, it does at least bang the futurity drum with some vigour: 'Development that meets the needs of the present without compromising the ability of future generations to meet their own needs'.

Speed and sustainability

Theoretically, every government under the sun is signed up to that overarching understanding of sustainable development. In reality, however, you will not see any particularly visible recognition of the

3 M. Crichton, *Prey* (New York: HarperCollins, 2002).
4 WCED (World Commission on Environment and Development), *Our Common Future* ('The Brundtland Report'; Oxford: Oxford University Press, 1987).

interests let alone the rights of future generations as we set about the serious business of having as good a time as we can before the whole house of cards comes crashing down around us. And doing so at full speed, with not the remotest concern for precaution. Today's fast-moving consumer societies in the rich world are now driven by an insatiable thirst for the new, for the next, for the neural numbing of life lived as a life-long shopping spree. And it's hard to ignore the fact that the emergence of today's increasingly sophisticated information and communication technologies has greatly exacerbated that consumerist frenzy.

I'm increasingly persuaded that speed may in itself be as much the enemy of a sustainable future as the reckless consumption that powers our global economy. They are, of course, closely linked. An essential attribute of our contemporary model of progress (in which consumption substitutes for quality of life or personal contentment) is that the faster you can do something, the better it must be. Eat faster, get the news faster, communicate faster, date faster, mate faster: 'life in the fast lane' is the aspiration of countless millions, regardless of the career crashes and life-wrecks that litter that particular lane. Jeremy Clarkson,[5] the high priest of speed for speed's sake, has a lot to answer for.

> Speed may in itself be as much the enemy of a sustainable future as the reckless consumption that powers our global economy

Far from improving the quality of our working lives, there is growing evidence that this progress is making us more miserable and more stressed. Money can buy us goods and services undreamed of in previous centuries, but it still can't buy us love or meaning, or at least not for long and not reliably. The pursuit of financial security is often associated with declining quality of life in much of the West, and the latest study from the ESCR (Economic and Social Research Council) on modern employment trends indicates a rising dissatisfaction with working life over the past decade.[6]

In *Britain on the Couch*, Oliver James summarised a mass of research data on the psychology of affluent societies, concluding that

5 Presenter of British consumer motoring television programme *Top Gear*.
6 R. Scase, *Britain towards 2010* (London: ESCR, 2004).

rates of depression, suicide and drug dependency are increasing in part because the competitive pressures of modern life combine to produce unhappy, tense and rancorous personalities. He goes on to assert that 'the closer a nation approximates the American model of a highly advanced and technologically developed form of capitalism, the greater the rate of mental illness amongst its citizens.'[7]

Why might that be? In the teeth of a generation of amoral relativists, the Chilean economist Manfred Max-Neef has argued that it's possible to identify a number of core human needs that are common to all cultures and all societies: subsistence, protection, affection, understanding, identity, creation, participation, leisure and freedom. What differs from culture to culture is the way in which we choose to satisfy those needs — the 'satisfiers' that we favour both at a societal and at an individual level. Our most pressing problem in the rich world is that we've lost the knack of distinguishing between genuine satisfiers and the kind of pseudo-satisfiers that leave the need unmet and our lives unfulfilled. Fast-lane living ranks high in any list of pseudo-satisfiers.[8]

But is it fair to single out today's information and communication technologies as a principal cause of this societal debacle? There's a simple test. On one side of your mind-map of how the world works, summon up (in a completely non-judgemental way) all the things that drive the wheels of commerce and material progress: efficiency, markets, productivity, growth, just-in-time delivery, doing deals, marketing, financial controls, taking risks, making money and so on. On the other, summon up (in an equally non-judgemental way) all the things that might be said to generate real quality of life: friends, children, food, music, having a good time, sport, hobbies, supportive and safe communities, gardening, reading and so on.

Self-evidently, today's new technologies are comprehensively implicated in speeding up the wheels of commerce in the pursuit of

7 O. James, *Britain on the Couch* (London: Century, 1997).
8 M.A. Max-Neef, 'Development and Human Needs', in P. Ekins and M.A. Max-Neef (eds.), *Real Life Economics: Understanding Wealth Creation* (New York: Routledge, 1992).

material progress. That is where they add the greatest value, and simultaneously set the dervishes of mass consumption and crude materialism whirling just that bit faster. At the same time, they do of course also impact on our pursuit of a better quality of life, especially in the fields of personal communication, entertainment and recreation. But the 'added value' they provide here is significantly less; the real enduring value on this side of the balance sheet still resides in the quality of our relationships rather than the virtuosity of the technologies we may use to support and enrich those relationships.

> The real enduring value resides in the quality of our relationships rather than the virtuosity of the technologies we may use to support and enrich those relationships

The evidence cited by Richard Layard at the London School of Economics suggests that this endless search for satisfaction through greater wealth (and thus we may assume accumulating the latest hi-tech gadget) has left the developed world no happier than we were 40 years ago.[9]

Let me put it more provocatively: the greater the enduring human value we derive from something, the less benefit today's technologies have to offer; the more something has to do with material consumption and spinning the wheels of ephemeral commerce, the greater the benefit such technologies may confer. If one subscribes to the view that the balance of our lives today (between commerce, consumption and materialism on the one hand, and culture, community, love and nature on the other) is profoundly distorted, then today's information and communication technologies must pick up their share of responsibility for that devastatingly destructive imbalance.

This is so unfashionable a view as to guarantee accusations of neo-Luddite, technophobic fuddyduddyism. But the contemporary debate about sustainability circles uneasily and intriguingly around the tension between dynamism and constancy. How, in a genuinely sustainable society, will we simultaneously nurture those aspects of

9   Prof. Lord R. Layard, 'Happiness: Has Social Science Got A Clue?' (Lecture 1, London School of Economics, 2003).

human nature that make us so uniquely powerful in evolutionary terms (a hunger for change, our insatiable curiosity, our pioneering spirit, our ability to manipulate nature, and so on), and those aspects of human nature that depend on a different rhythm — cultural continuity, stable communities, time for reflection and spiritual devotion, an enduring, symbiotic relationship with the natural world — with its very different cycles, seasons and time frames? In truth, I cannot see today's technologies as balancing agents in managing that tension. Indeed, I believe they are deeply implicated in the lies and illusions involved in propagating the 'benefits' of living faster, all too seductively selling the untruth that speed is a satisfactory surrogate for genuine contentment and quality of life.

Governments and big business need us all out there consuming diligently and patriotically

The reality is that we are completely trapped in this utterly perverse model of consumption-driven progress. Whether it makes people more or less contented is irrelevant: governments and big business need us all out there consuming diligently and patriotically (remember George Bush's call to the American people after 9/11 to get out there and flex their credit cards?) just to keep the wheels of the economy spinning as fast as possible.

The population paradox

In that structurally dysfunctional context, a declining population becomes a major headache for politicians. Fewer people means both less consumption and lower tax receipts. Which means that everyone in that reduced population is just going to have to consume even more diligently! While it's undoubtedly true that there are some genuine problems arising out of this 'demographic time bomb' (i.e. an ever lower ratio of the young and employed to the old and retired), particularly in terms of securing a decent state pension for older people, it seems to me these problems are often flagrantly exaggerated to reinforce the very model of economic progress that is bringing the planet to its knees in the first place.

Just four things would go a long way to diffusing this particular time bomb here in the UK: distributing available work more equitably, securing in the process a better work–life balance for millions of workaholics; raising the pension age to 70 for all men and women alike; radically re-engineering tax incentives to promote saving rather than consuming; and maintaining an intelligent, rigorous immigration policy to ensure no more than a small net increase of those coming in every year over those going out.

It's all a question of timing. That last policy, for instance, would have the direct consequence of slowing the necessary reduction in the overall UK population. If that goal was our *sole* policy priority, then we would need to ensure a strict balance between inward and outward migration, or even tougher controls to reduce or eliminate immigration altogether. That, for me, is unacceptable. But the clear implication of taking a more balanced approach (with the reduction in UK population spread over a longer period of time) is that we need to be even more focused in the short term on policies to further reduce average fertility (among all citizens, established and new arrivals), dramatically reduce unwanted pregnancies (particularly among teenagers), and disincentivise families of more than two children.

Not uncontroversial, it has to be said. Which may well explain why calls to the UK government from organisations such as the Optimum Population Trust for the incorporation of a formal Sustainable Population Policy (and target) in the new Sustainable Development Strategy are likely to fall on resolutely deaf ears. Politicians just don't do population if they can possibly avoid it, ensuring that one of the most crucial long-term aspects of sustainable development is simply ignored.

With one bizarre exception: a growing band of politicians and religious leaders have started to advance stridently pro-natalist views in those countries where population has been declining over some time. In the September 2004 issue of *Free Inquiry*, Tom Flynn sums up the current situation:

Tragically, the opportunity presented by falling birthrates is being misperceived as a crisis. Frantic leaders call for a return to the geometrical birthrates of old. It's no surprise that the Pope hectored Italians to 'reverse "the crisis of their birthrate" by having more babies'. It is more remarkable that Sweden and Italy offer new tax breaks for parents. Towns from Europe to French Canada subsidize large families. Convinced that citizens aren't having enough sex, Singapore's government runs matchmaking services. State-controlled media urge, 'Let's Get on the Love Wagon'. But the record for bluntness goes to Australia's Treasurer, Peter Costello. Stumping for a budget that would pay a bounty of A$2,000 for each baby born, he called for three-child families and exhorted his countrymen (and women) to 'go home and do your patriotic duty tonight'.[10]

Consuming and breeding: the hallmarks of perverted patriotism in the 21st century. The challenge of learning to live sustainably on planet Earth over the next 50 years (which is probably a massive overestimate of the amount of time available to us to pull off that particular transition without traumatic social and environmental disruption) is tough enough without political mountebanks of that kind stacking the odds against us. But we are as yet still so far from understanding humankind's role in both nature and evolution that it's hard to imagine how things are going to start getting better without first getting a great deal worse.

> We are as yet still so far from understanding humankind's role in both nature and evolution that it's hard to imagine how things are going to start getting better without first getting a great deal worse

10   T. Flynn in *Free Inquiry* 24.5 (September 2004).

Calendars, creation and the end of time

It is difficult to imagine life without calendars. Daily, monthly or yearly, they allow us to plan into the future and to measure the past. They have existed for thousands of years; they are a cornerstone of civilisation as we know it. Even Robinson Crusoe kept one when stranded on his desert island.

The calendar we use today is a mishmash of overlapping cultural systems. Its structure follows that of the **Gregorian calendar**, first instituted in 1582 by Pope Gregory XIII, which England adopted in 1752. Some of the units into which we divide time have a close relationship with nature while some are arbitrary or historical. Christianity gives us Christmas and Whitsun; Mayday and Hallowe'en are relics of pagan rituals; we observe dates of political significance such as 11 November and the Queen's birthday; and our summer holidays are based around farming timetables. Our days follow the rotation of the Earth; years follow the cycle of the seasons, the equinoxes and solstices, the Earth's orbit around the Sun.

Thousands of years ago, by observing the movements of the Sun and Moon, ancient civilisations were able to devise complex methods of timekeeping. The **Maya** people of ancient Mesoamerica acquired an excellent knowledge of astronomy, allowing them to develop one of the world's most sophisticated calendar systems. The *tzolkin*, their sacred 260-day calendar, was based on lunar cycles, while a 365-day calendar, the *haab*, was based on the solar year. Each day had both a lunar and solar designation made up of names and numbers; it took 52 years for the sequence to repeat itself, in a cycle called the Calendar Round.

The Maya lent great significance to the 'long count' of time. Instead of three figures for day, month and year, their dates used five figures, starting with the number for the *baktun* — a cycle of 144,000 days, or twenty times twenty 360-day *tun* cycles. According to the widely accepted Goodman–Martinez–Thompson correlation with the Gregorian calendar in use today, the first day of the Long Count, [0.0.0.0.0] in the Maya dating system, corresponds to 11 August 3114 BC. The current deep time cycle of 13 *baktun* lasting 1,872,000 days ends on the Winter Solstice (21 December) in AD 2012. This does not necessarily mark a permanent end, as the Maya believed that the world had been destroyed and created several times.

The **Ancient Egyptians** are also known for the complexity of their calendar. They studied the tides of the Nile, which, combined with astronomy, gave them a unprecedented understanding of the rhythms for nature, and the knowledge that a year has 365 $\frac{1}{4}$ days, not 365. However the priests were reluctant to make any changes, so this wasn't reflected in the calendar system until Julius Caesar instituted the **Julian calendar** in 48 BC. Before his reforms, the calendar was so far off course that people would celebrate the changing of the seasons at the wrong time of year. To get things back into place the year 46 BC lasted 445 days; two months with 33 and 34 days were inserted between November and December in what was called *annus confusionis*.

Like Caesar, whose Julian calendar determined the running of the Roman Empire, political groups throughout history have used calendar systems as an extension of their power. Both the **French and the Russian** revolutions were accompanied by attempts to change the calendar system. After toppling the monarchy, the National Convention, or French Parliament, decided to introduce a completely new calendar embodying the spirit of the revolution and expunging all traces of God and King. Year 1 began on 22 September 1792, the day the Republic was proclaimed.

French dramatist Fabre d'Églantine created new names for the twelve new 30-day months, and for every day of the year, derived from the names of fauna, flora, minerals and agricultural tools. The last five days of the year were set aside for a festival of *Sansculottides*, after the *sans-culottes* (without knee-breeches) revolutionaries. Though the revolution's motto of 'liberté, egalité, fraternité' survives to this day, its calendar lasted little more than a decade, abolished by Napoleon Bonaparte. On 1 January 1806, France adopted the Gregorian calendar once more.

Russia, too, adopted a new calendar after its revolution, which was meant to increase productivity. Ironically, the 'Eternal Calendar', as it was known, lasted only a few years. Its five-day calendar week and staggered rest days were abandoned in 1932, and the Gregorian calendar was brought back in 1934.

The Romans originally dated their calendar *ab urbe condita*, from the founding of Rome. Then in AD 525 a Scythian monk, Dionysius Exiguus, devised a new system of dating from the year Christ was born or conceived. The **Anno Domini Nostri Jesu Christi** system[11]

---

11   Latin: 'In the Year of Our Lord Jesus Christ'.

(AD for short) was promoted by Bede and gradually spread through-out the Christian world. There is some controversy over the true year of Jesus Christ's birth, however, and it has been suggested that 'year 1' should actually be several years earlier.

Anxiety, apocalyptic prophecies and civil unrest were reported in the approach to the years 1000 and 2000. In both cases certain fol-lowers of Christianity believed that the millennium would mark the beginning of end-times preceding the second coming of Jesus Christ, as described in the Book of Revelation.

Unlike the Gregorian and Islamic calendars, based respectively on the Moon and the Sun, the **Hebrew calendar** is lunisolar, combining both systems. Hebrew years follow the Sun while months follow the lunar cycle: days begin at nightfall and each new month begins when the crescent of the moon first becomes visible in the night sky. The New Year, Rosh Hashana, takes place in autumn. Traditionally, Juda-ism dates creation back to the first year of the Hebrew calendar, on the 25th of the month of Elul, almost 5,800 years ago.

The **Chinese calendar** is also based on a combination of Moon and Sun movements, which is why the New Year isn't on the same date every year. The moon cycle lasts about 29.5 days. In order to 'catch up' with the solar calendar the Chinese insert an extra month once every few years (seven years out of a 19-year cycle) — in the same way adding an extra day on leap year. According to many, the Chinese calendar dates back to what most of us would call 2637 CE (Common Era), making 9 February 2006 in the Gregorian Calendar the first day of 4642 in the Chinese calendar, which is, incidentally, the year of the rooster. Like Julius Caesar, in ancient times the Chinese Empire used its calendar as a symbol of power: conquered nations were said to have 'received the calendar'.

The latest attempt to change the time measuring system was by the Swiss watch company, Swatch, which in 1998 launched **Internet Time** — inventing a new time unit and a new way of telling what time it is. Internet Time divides the day into 1000 'beats', each equivalent to 1 minute and 26.4 seconds. In true Internet spirit the current Internet Time is the same globally, and the proposed advantage is easier time co-ordination over different time zones. Some Swatch watches come with beats as well as minutes and hours, but other than that it is difficult to see its impact. Internet Time looks like a flop by a global corporation with megalomania.

Britt Jorgensen and Mireille Kaiser

4

Living time

Jay Griffiths

The Karen always know the time. While living with them for six months it became clear to me that the only person with a watch and the only person who could never tell the time was, well, myself. To the Karen people of Thailand, the forest over the course of a day supplied a symphony of time, provided you knew the score. The morning held simplicity in its damp air, unlike the evening's denser wetness when steam and smoke thickened the air. Backlit by sun, a huge waxy banana leaf at noon became green-gold stained glass, cathedralising time. Barely one of my hours later, it was just a matt, bottle-green leaf, useful verdure, a plate for rice, a food-wrapper. Birds sang differently at different hours and, while the soloists of life are always with us, the whole orchestra of the forest altered, shifting with the sun's day, all the noisy relations between birds, animals and insects making chords of time played in all the instrumental interactions. The time on my wristwatch, squeaking its repeated identical numbers, seemed a thin, thin reedy peep of a thing by comparison.

The Karen always know where they are and when they are, how far they are from sunset or home: for time and distance are connected in the Karen language: *d'yi ba* — soon — means, literally, 'not far away'. Sunset, therefore, could be expressed as 'three kilometres away', because the only way of travelling is to walk, which takes a known length of time.

There was a similar indivisibility between social time and action. I — eventually — learned not to ask 'When will such-and-such happen?' When will a wedding, for instance, take place? For the no-answer-possible smiles replied: It happens when it happens — the doing of a thing and its timing are indivisible, the action is not jostled into the hour, but the hour becomes the action and the action becomes the hour. What to me was a distinction between the hour and the act was, to them, tautology, half artificial and half daft.

In the modern Euro-American culture, time is a dead thing, a disembodied ghost no longer embodied in nature

Dead versus living time

In the modern Euro-American culture, time is a dead thing, a disembodied ghost no longer embodied in nature; the moment struck dumb by the striking clock, the deadening character of routines, schedules and endlessly counted and accounted time. By contrast, in most cultures, for most of history, time has one supremely different quality; time is alive. And is lived as such. The Euro-American image of time is a machine, a factory assembly line chucking out identical hours, each unremarked and indistinguishable. Worse than that, it has insisted that *its* time is *the* time, and that indigenous peoples all over the world lack a 'proper' sense of time. It is not a lack. Rather they have cultivated a far more subtle and sensitive relationship to time and timing.

Indigenous people regard time as inseparable from nature, in contrast to the numbering abstractions of Western time disembedded from nature. The Leco people of Bolivia have tree calendars, the

U'wa of Colombia have insect clocks 'which whistle on the U'wa hour' and the Kaluli people of Papua New Guinea have a clock of birds. The Kaluli say that the early morning calls of the brown oriole and New Guinea friarbird and the hooded butcher-bird tell the children to wake up, and later in the day, the birds' afternoon calls tell the children to return to their families.

Indigenous people regard time as inseparable from nature, in contrast to the numbering abstractions of Western time disembedded from nature

The San Bushmen of the Kalahari would never schedule when to hunt but would read and assess animal behaviour and choose a 'right' time spontaneously, 'waiting for the moment to be lucky'. For the Ilongot people of the Philippines and for every indigenous group I have ever known or read about, timing in social interactions is indeterminate, unpredictable, demanding flexibility, fluidity and quick co-ordination. It is a graceful, alert way to live, demanding acute skills of psychology, as well as keen observation of weather and animal movements. Hunter-gatherer time is a series of unique moments, confluences of a hundred streams, a thousand interconnecting factors, including a person's mood, a shift in wind direction, knowledge of a cubbing season, a sight of fresh tracks. Scheduling or planning would destroy the necessary elusiveness of this subtle sense of timing, and would kill stone dead the exquisite sense that time is alive. Hunter-gatherer time is local to the spit of rain, a shadow in the bush.

Naming time

Sure-yani Eduardo Poroso, a leader of the indigenous Leco-Aguachile people in Bolivia, comments:

> The Leco people don't use the Roman calendar but nature's calendar. You can't use the Roman calendar to know when to fish. It's imposed on us for registering births and for baptisms, but it doesn't function for use in the forests. Bats tell us when to fish; when they fly close over the water.

His own name is an indigenous time-referent; Sure-yani is the name of a 'calendar-tree' which lives for 300 years and is, he says, 'how we measure time'.

Most societies name certain times with terms of nature. In parts of Indonesia, two of the waxing nights of the moon are called the 'little pig moon' and the 'big pig moon'; the nights the Roman calendar would call the eleventh and twelfth nights of the moon's growing. To the Khant people in northern Siberia one month is called the Naked-Tree month followed by the Pedestrian month when people must journey not on horseback but on foot — gingerly — over ice. The month of Crows (also called Wind month) comes before the Spawning month, then follow the Pine-Sapwood month, the Birch-Sapwood month and the Salmon-Weir month. The variations of landscape or animal appearances characterise both time and locale, in vivid contrast to Western time measurement, its invariance characterising neither time nor place. Across the world the nature of each moon-month is characterised and, through each people's names for the months, you can 'see' the specific landscape they inhabit.

All over the world, people have celebrated festival times within nature: Native Americans celebrating the first dew of spring or the first caterpillar of the year; the Kayapo of Brazil having festivals for maize and manioc seasons and for the hunting seasons for turtle, tapir and anteater. In the Micmac culture of North America, feasts celebrate events of nature like the new year which happens at the new moon when the creek is frozen, not a disembodied hollow of number, 01/01. There is in the Micmac language, and in most of the Algonquin languages, no word for 'time' in an absolute sense. There are words for day, night, sunset, sunrise, a year and moon, but no 'time' as numbered measurement.

There is in the Micmac language, and in most of the Algonquin languages, no word for 'time' in an absolute sense

The Hopi of northeast Arizona express the indivisibility of time and nature in their very language. Time, to them, is not an abstract mechanical thing, but an intrinsic process of nature. Time happens in the moment of the biting of a scorpion, or in the months of a crocodile growing up; time is identical to the maturing of the baby whippoorwill. It is the process, or duration, of nature. One indigenous language in Madagascar, tying time directly to nature, refers to a moment as 'in the frying of a locust'. The Cross River peoples in Nigeria, describing the exact length of time it took for one man to die, a period we would call less than 15 minutes, said that 'he died in less than the time in which maize is not yet completely roasted'. The English language does retain a shred of nature in its terms for time-keeping; doing something in 'two shakes of a lamb's tail', but it is rare. 'Pissing-while' is recorded in the *Oxford English Dictionary*, which certainly ties time to the call of nature and does so in a delightful if arbitrary stream of thought.

The end of time?

Though the Euro-American world assumes that the linearity of time is a universal truth, it is a rare and recent idea. Instead, most societies throughout history have perceived time to be cyclical: the Hopi image of time is a self-contained wheel. In Hindu thought, time moves in the unimaginably long cycles of the Kalpas. In the *aions* of the ancient Greeks, eternity wheeled round over and over again. Aristotle said: 'For even time itself is thought to be a circle.' Some societies recognise neither line nor circle: the Navajo speak of 'pulses' of time, while ancient Indian philosophy teaches of a 'vessel above time' always overflowing, a non-geometric image of lovely, liquid eternity. If you see time as cyclic, you have an immediate and strong image of sustainable time; running its hours around the day, coursing in years that are endless, sustained and profoundly lifeful.

Though the Euro-American world assumes that the linearity of time is a universal truth, it is a rare and recent idea

By contrast, the linear time introduced by Judaism, and the end-stopped time that the Christian era introduced, accustoms its believers to finality, the end of the world. The cyclical time to which nature-based cultures adhered was furiously attacked by the Christian church: St Augustine wrote that the history of the universe is 'single, irreversible, unrepeatable, rectilinear', and 'it is only through the sound doctrine of a rectilinear course that we can escape from I know not what false cycles discovered by false and deceitful sages'. The vital and self-recreating time of cyclic eternity has been represented, all over the world, by the snake, sloughing off its old skin and beginning again. Judaeo-Christianity turned the cyclic snake into a figure of hate.

Christianity was the main force behind the globalisation of time and the crushing of various times across the world; Anno Domini was invented in 525 by Dionysius Exiguous. Some five years later came a hugely influential and novel attitude to time, the Rule of Saint Benedict. There were schedules and timetables for everything: prayer, work and sleep. It brought in a coercive time discipline, in which human nature was bent to the clock. Idleness, that impish, happy spirit, was decreed 'the enemy of the soul'. Time was strictly regulated into the Offices of hours. And then the Christian missionaries imposed their time, and their time-values, on cultures across the world. When missionaries arrived among the Algonquin peoples of North America, the Algonquin called clock-time 'Captain Clock' because it seemed to command every act for the Christians. In *Gulliver's Travels*, too, the Lilliputians observed that Gulliver's God was his watch.

The Algonquin called clock-time 'Captain Clock' because it seemed to command every act for the Christians

Once, you could say that time was so local that, for every *genius loci*, a spirit of specific place, there was a *genius temporis*, a spirit of specific time; but timekeeping of the dominant Euro-American culture has brought universal homogeneity and standardisation. The clock-time of modernity, applicable anywhere and imposed all over the world, is one aspect of globalisation, the globalisation of time. The Roman clock and calendar became *the* time and *the* date, as Western — specifically British — time was imposed all over the world. More perniciously and highly politically, the time-values of this one culture have been forced onto all. Efficiency, punctuality, speed; the idea that the past is dead and, perhaps most objectionable of all, the West's idea of progress, have been imposed on cultures whose own wisdom preferred sensitivity, a living past and sustainability. What I think of as the 'politics of time' is a matter of cultural imperialism, and racial and religious subjection.

Time and power

The march towards mono-time can be shown obviously in law; from 1840, railways in Britain required a standard time and, in 1880, London Time was decreed by law to be the time for the whole country. In 1883, the US standardised timekeeping for railways; in 1884, Greenwich was made the zero meridian and the global day of 24 hours began, a day made of the same hours, irrespective of local cultures' clocks.

But there are less obvious methods than law by which the world's myriad and diverse times came to be dominated by one. Keep an eye on the clock. Keep the other eye on the seas, which many cultures have understood as intimately connected to time. In the tides' ebb and flow, the moment is critical and, while at the tidal shoreline the sea represents the now of events, the paradox of the ocean is that in its depths it is a symbol of eternity. Byron called the sea 'the image of Eternity'.[1] To Western scientists the sea is the source of life. In Taoist thought, the ocean is equated with the Tao, primordial and inexhaustible source. Jain thought describes eternity as an 'ocean of years'. And Otis Redding picked the right place — 'sittin' on the dock of the bay, wastin' time'[2] — for the sea is creator of endless hours of time. In English, the word 'tide' can be used for both the sea's tides and certain times — noontide, eventide, Whitsuntide or Eastertide. The word 'tide' is etymologically related to the word 'time' in Old English. 'Current' refers to both time and tide.

1 Lord Byron, *Childe Harold*, canto IV, st. 183 (1818).
2 '(Sittin' On) The Dock Of The Bay', written by Otis Redding and Steve Cropper, recorded by Otis Redding, 1967.

In an historical sense, time and the sea are linked: the great breakthrough of timekeeping — inventing clockwork of sufficient accuracy to discover longitude — was made in order to gain control of the seas; which led to Britons ruling the waves and, through this, ruling both empires of land and empires of time, for it was due to Britain's maritime supremacy that Greenwich was accepted world-wide as the zero meridian. Ruling the seas meant ruling the standard of world-time — the time that is so highly political and so seldom recognised as such. The most accurate clocks were kept at the Greenwich Royal Observatory, the centre of this maritime nation and the centre of empire. The chief clock at Greenwich in 1852 was called the 'master' clock and it sent out signals to 'slave' clocks in Greenwich which sent further signals to other 'slave' clocks at London Bridge. The same language is still in use for clocks in both Britain and in Washington where the Naval Observatory has its master clock facility and slave clocks are still designed.

Potentates, princes and priests, all hypnotised by hopes of hegemony, have always stood on the borders of space and looked at time — they have come, they have seen, they have conquered, for time is a kingdom, a power and a glory. Pol Pot declared 1975 Year Zero. Hitler defined his political ambitions through time: the Third Reich was to last 1,000 years. Stalin tried to cancel Sunday and, to demonstrate Moscow's power over other parts of the Soviet empire, Moscow Time was decreed the standard of Communist time. When ancient China had colonised some new region, the phrase they used to describe this act was at once sinister and telling — the people of the new territory had 'received the calendar'.

Since time has long been a locus of power, it is unsurprising that it has been a focus for revolutionaries and protesters. Trades unions took on first the abuse of time. In the French Revolution, 1792 was designated Year One, so time past was to be guillotined from time present. Similarly in the British road protests of the 1990s, the hilarious 'Trolls' of Trollheim, at Fairmile in Devon wrote their manifesto of the Independent Free State of Trollheim:

> We do not recognise history, patriarchy, matriarchy, poli-
> tics, communists, fascists or lollipop men/ladies . . . We
> have a hierarchy based on dog worship. Our currency is to
> be based on the quag barter system. We do not recognise
> the Gregorian calendar: by doing so this day shall be known
> as One . . .

The Zapatistas have altered their clocks so that they run an hour
ahead of Mexico City and, more deeply, they have insisted that their
kind of time was not that of the Mexican government. The Zapa-
tista leaders took their orders from the peasants — a process both
very slow and completely unschedulable. They commented, 'We use
time, not the clock. That is what the government doesn't understand.'

As a consequence of its linear view of time, the Euro-American
mind sees the past as 'dead and gone' while many indigenous views
see the past as profoundly 'alive' — in the land. When the Arakmbut
people of the Peruvian Amazon think of the past, they think of
Earth: 'Without the knowledge of history, the
land has no meaning and without the land neither the Arakmbut history nor the culture has any
meaning.' The Quechua concept *pacha* means
both 'land' and 'time'. Sure-yani Poroso comments: 'Land is history for us. It is our past, our
present and our future.' The Australian ancestors
'live' in spite of death: they disappear but have not
died. They become immanent in the land. 'History
comes up from the land,' says Aboriginal author Herb Wharton.
The land is animated with the past — filled with sacred energy
according to indigenous philosophy — so the past still exists, an
opaque modality of time-present. Life runs its courses underground. But there's more. The inherently differing notions of the
past have direct — and contemporary — political consequences. The
underground past is a source of sacred energy to indigenous peoples but merely a source of literal energy, fuel, to the Euro-American
mind. So mining companies devastate indigenous territories, which
is sacrilege to their earth-based way of thinking.

The land is animated with the past — filled with sacred energy according to indigenous philosophy — so the past still exists, an opaque modality of time-present

The U'wa people in the cloud-forests of Colombia consider the 'past' of nature in the oil reserves underground as the 'blood of the earth' and say the land is 'alive' with oil. To take out the oil is to kill the land and themselves, for without the land, they say, they 'are not'. The oil from U'wa lands would, it was thought, sustain global energy demands for a maximum of three months: thousands of years of sacred indigenous past gone to fuel 90 days of the Western present.

At dawn, the U'wa daily sing the world into creation by reciting their myths and place names. Through sustaining nature, they believe, they sustain time itself — sustainability is associated with sheer life. At dawn, too, the skies are full of birdsong: whistling, riffing, cooing, calling, croaking and trumpeting, the birds' fluting songs echo off the clouds. When the birds streel across the skies, they chant the names of the areas they fly over and, by doing so, they create places; or so the U'wa believe.

But these cloud-forests have been invaded by modern *conquistadors*, the oil companies, so that Western culture can fuel its cars, the powerful symbols and engines of its progress. The U'wa say: 'We are seeking an explanation for this "progress" that goes against life. We are demanding that this kind of progress stop, that oil exploitation in the heart of the Earth is halted, that bleeding of the Earth stop.'

Progress and sustainability

Progress is a specific idea; money-oriented, technologically biased and racist in its history and its effects, but it pretends to a universality, so that all peoples must be made to define and embrace progress in exactly the same way. The U'wa don't. They know their progress is not served by oil companies destroying their land for Euro-American progress. 'Your progress is not our progress' is a phrase I've heard all over the world, from Native Americans and

Amazonian people to Maori and others. Progress pretends to be an absolute good because it is defined by those it serves well; the rich, the politically powerful, all types of colonialists and ideologues. Ask those whom it serves badly and they will tell you that the engines of 'progress' have justified the destruction of lands and peoples of the land, from racism to land thefts, from pollution to extermination. 'nothing regresses like progress' wrote ee cummings.

Progress is a one-word ideology and one that has suited both the Soviet Marxist world-view and also that of neo-liberal multinationals and global free-marketeers. The latter push it as an inevitable, neutral 'good', silencing its highly political character. Devotees of progress can be sneeringly contemptuous of anyone who would dare speak against it and those who stand in the way of progress are called ridiculous, backwards and reactionary. Or worse. Sure-yani Poroso was tortured by the authorities in Bolivia for campaigning for his tribe, against neo-liberalism, land thefts, water privatisation and the TransAmazonica highway. 'For doing this I was called anti-progress. When I was being tortured, they said "You are a shit. This is what you get for fighting progress."' When the word on a torturer's lips is 'progress' you see the vicious side of its ideology.

'Your progress is not our progress' is a phrase I've heard all over the world

It seems appropriate that, all over the world, the finest critics of so-called progress are people of the land. For, as in the first colonial era of state colonialism, it was the indigenous races who suffered most from the European desire for 'progress', so in this era of corporate colonialism, it is people of the land who suffer from modernity's progress away from nature.

It's a clash of two cultures and two mind-sets: one characterised by off-ground ideologies and by the idea of the temporary — the flashy mobility built in to Euro-American cultures; the other is based on permanence and a love of place.

According to Western culture, speed is god. 'Be fast or be last'; 'You snooze, you lose'; we have speed dating, speed funerals, fast food, fast trains, fast drugs (coffee, Prozac and speed). Many societies loathe this petty god. To the Kabyle people of Algeria (and indeed many others from the Inuit to Aboriginal Australians and Apache people), speed for its own sake is indecorous and demonically over-competitive. Kabyle people refer to the clock as the 'devil's mill'.

Amilton Lopez is a Kaiowa chief, of the Guarani-Kaiowa peoples of Mato Grosso do Sul, in Brazil. He came to London while I was researching my book, *Pip Pip*, to talk about 'progress'. Land that used to belong to his tribe is now being taken by cattle ranchers for the progress of burger culture — and his tribe is devastated. As their land is taken away, suicides are increasing, particularly among young girls. There were 43 suicides in 1995 alone. 'People are killing themselves because they have literally no more space to live in. Everything is finished. The forest is finished so our culture is finished because our culture is in the forest, in nature, in the environment.' In one Kaiowa area, people have threatened, like the U'wa, to commit collective suicide if their land is taken from them. 'We *are* the land,' he says, willing me to understand that profound simplicity, and he can't say any more: the words choke him.

'People are killing themselves because they have literally no more space to live in. Everything is finished'

No one can live far enough away from Euro-American progress for, in a conceptual sense, there is no place left on Earth that this 'progress' does not overshadow, this progress which is such an enemy of place, of land, enemy of the people of the land, enemy of Earth itself. Only by dying can they move far enough away.

In English, the very word 'progress' has undeserved positive associations; similarly the word 'sustainability' has equally undeserved negative associations. 'Progress' seems to conjure the vigour of motion, the fire-fare-forwards of life itself. 'Sustainability' seems to weigh in with the burden of a heavy stasis, a life half-lived and a death half-died; 'sustainability' has all the dirgeful effort of a worthy cause and none of the dynamite of 'progress'. But the opposite is true. Progress, along the trajectory Euro-American culture is now on, is a one-word lie; it is neither the travel nor the arrival, but the ultimate ending; not the flame of thought, but a bonfire of humanity: the vaunted 'progress' of cars and unlimited plane travel leading to global warming and millions of environmental refugees. This so-called progress is a politics that tends towards death. Sustainability, on the other hand, is where the life lies, where time touches eternity, the time of the natural world, of ice and melt, of the seas' times and tides. Both sustainability and progress need to be redefined and reclaimed. In order to do this, Western culture needs to listen to indigenous peoples because, in their ideas of cyclical time, time is constantly restored, nature sustained and sustaining. These are the very ideas the world needs most. The fire of the human mind is needed, for fireworks and for warmth, not arson; for creating a progress of magnificence; for protecting the myriad splendour of diverse cultures, languages and species, in an august roar to increasing, noisy, earthy vitality. Sustainability needs to be redescribed, as fleet, hopeful, sensitive and passionate — a vision of the future that looks ahead hundreds of thousands of years. If sustainability means anything, it means constantly self-renewing time, generation and regeneration, generous to overflowing. Sustainability is not an issue but a synonym. For *life*.

The anthropology of time

In *The Silent Language*,[3] anthropologist Edward T. Hall examined the way time is processed and structured by different cultural groups. Hall divided time into two categories: monochronic and polychronic. Monochronic time, like money, can be saved or spent, whereas polychronic time is more flexible, characterised by several activities taking place at once.

Robert Graham and Alma Owen further developed Hall's idea into **three anthropological models of time**.

The **linear–separable** model dominates the Western view of time. In this way of thinking, past, present and future are distinct entities which can be broken down into units such as minutes and centuries. Long-term planning is considered normal, frequently using steps and procedures such as you might find in a flowchart or business plan. Cultures that favour linear time-processing value speed of preparation; hence the popularity of time-saving products such as microwaves, canned soup and faster computer processors in Europe, North America and Japan. 'It was in western Europe that the mechanical clock first appeared and with it a new type of civilization based on the measurement of time.'[4]

The **procedural–traditional** model of time sees events and procedures as being more important than the time they take, as when a scientist conducts research, or an artist creates a painting. Being on time matters less than timeliness, and doing things well. As well as being prominent among several tribes of Native Americans and Alaskan Eskimos, the procedural model is influential in many Latin countries, where a 45-minute wait for an appointment would be on the short end of the 'waiting scale', according to Hall. In Japan or the US, on the other hand, such a delay would be considered insulting.

The **circular–traditional** model sees time as having a rhythmic pattern. There are beginnings and ends but no certainty of seeing change, and no discrete units of past, present or future. To the Inuit of Canada's Baffin Island, the same word — *uvatiarru* — means both 'in the distant past' and 'in the distant future'. This view of time is often associated with life in poor or less technologically developed

---

3   E.T. Hall, *The Silent Language* (Greenwich, CT: Fawcett, 1959).
4   G.J. Whitrow, *Time in History* (Oxford: Oxford University Press, 1988): 96.

countries, where there may be little variation from day to day. Ancient Egyptian farmers, for instance, believed that in the afterlife they would continue to live as they had in their previous existence. This way of thinking is also associated with a much longer view of time than the linear model. As Hall writes, 'The future for us is the foreseeable future,' whereas South Asians feel 'that it is perfectly realistic to think of a "long time" in terms of thousands of years'. For this reason businesses in Asian countries may be planned over several decades, not just a few years. An important scripture of Taoism, the *Chuang Tzu* (4th century BC), sums up the circular model of time when it says: 'There is existence without limitation; there is continuity without a starting point.'

While they may favour one model over another, different people and different cultures often use several models of time concurrently, as in Western Europe where all three ways of thinking are present. And some cultures have a concept of time that is completely different to ours — in *The Silent Language* Hall also examines cultures that have no sense of past or future, and no words for these concepts.

The model of time favoured by a culture or individual has wide-ranging effects: for instance, on consumer preferences. In South America, fast-food chains such as Wimpy, KFC and McDonald's have made less of an impact than in countries that place a high value on speed. As French economist Michel Chevalier wrote in the 1830s, the 'American has an exaggerated estimate of the value of time and is always in a terrible hurry.'[5]

Britt Jorgensen and Mireille Kaiser

5 M. Chevalier, *Atlas: Histoire et description des voies de communication aux États-Unis et des travaux d'art qui en dépendent* (Paris, 1840).

5

The arrival of time politics*

Geoff Mulgan

For several decades it has been apparent that the old ways of managing time are fast disappearing. Fixed jobs, shared rhythms of shopping and leisure, marriage, work and retirement — all are on the way out, albeit slowly. Public policy, after lagging behind social change, has started to catch up with new rights to work flexibly that have captured the public imagination far more than anyone expected, rights to parental leave, and support for childcare that is for many the prerequisite for more flexible working. The advent of 24-hour health advice from NHS Direct, postal voting, online tax returns and of ministers who resign to spend time with their family and actually mean it, are all symptoms of this shift.

Three interrelated drivers of change have made the running. The first is technology, which is transforming the nature of economic life with vastly greater flows of information and much tighter control over time. The second is the continuing rise of a culture of

*   This essay draws on Geoff Mulgan's Jon Lopagetui lecture on time at Kingston University and on the Demos collection, *The Time Squeeze* (London: Demos, 1995).

choice and freedom which, among other things, has encouraged women to take control over their own lives and reject the old domestic division of labour. The third is economics: the simple fact that an hour's work is now worth 25 times what it was in 1830 brings with it the sense of time as a much more valuable resource to be managed, planned and used more intensively.

The effects of these forces for change can be discerned in the characteristic language of the contemporary era: just-in-time production and multi-tasking computers, 24-hour shopping and video-on-demand, time-share holidays and annualised working hours, late-night shopping and online learning, channel surfing and asynchronous email. Together they offer an extraordinary promise of much greater personal control over time. But, as always in history, things are not so clear-cut. While one foot steps forward another steps back. So, although lifetime working hours have fallen by over 40% over the last century,[1] a large majority are suffering from stress because of the rising intensity of work and leisure.

> An hour's work is now worth 25 times what it was in 1830

Ironically, both the overworked and the unemployed share the sense that their position is involuntary. Most adults have a sense of time being squeezed as they spend longer driving children to school, getting to the shops, even filling out tax returns, and as the world around them changes with bewildering speed (90% of new goods are no longer on the market two years after their launch).

But, even though many people are desperate to find more balance between work and other parts of life, their families and their enthusiasms, when people do get more 'free' time, not everyone knows how to cope. Retirement often brings illness and depression; the unemployed suffer the worst unhappiness of any group; a majority would work even if they

> An hour in 1990 is five times as productive as it was in 1830

1 Bruce Williams estimated a decline from 154,000 hours in 1881 to 88,000 in 1981 (B. Williams, *Shorter Hours, Increased Employment* [paper for OECD, 1984]).

didn't need the money; and few find their leisure time very satisfying.

This paradox is at the heart of the time issue. To explain it we need to go back to some basic principles — principles that can help us distinguish good and bad uses of time.

The first is a principle of choice. Wherever possible it should be a basic principle in a democratic society that people can determine, so far as possible, how they use their time. The second is a qualitative principle. Good uses of time are those that give enjoyment, that develop our potential, that leave something useful behind. Bad ones by contrast are inert, useless, mindless, unmemorable.

Together these two principles help to give us a better perspective not only on the present, but also on the past and future. For, by their standard, human history has not been a straightforward progress. The pre-agrarian hunter-gatherers spent only 15 hours each week engaged in work — work which was often more demanding of intelligence than most work today. By contrast, the back-breaking agricultural and repetitive factory jobs that succeeded them did little or nothing to use or develop potential.

What can we learn from the past? In pre-industrial societies, time is close to nature. Social life is ordered by the rhythms of the seasons, of day and night. People understand time cyclically, and in terms of key rites of passage

Lifetime working hours have fallen by over 40% over the last century

through life, moments that were often experienced communally — birth, the transition from childhood to adulthood, from education to working life, from work to old age. In such societies time is rarely measured; instead it is present subjectively rather than objectively. It is local, based on slow and steady rhythms, and the main role of policy is to regulate the festivals of saints' days, or seasonal celebrations, and the rhythms of the harvest (still reflected to this day in some of the techniques of accounting).

In the industrial era, as Lewis Mumford pointed out,[2] time takes on the character of the clock. It becomes mechanical, regular, re-

2   L. Mumford, *The Lewis Mumford Reader* (ed. Donald L. Miller; Athens, GA: University of Georgia Press, 1995),

moved from nature. It comes to be seen as a resource to be managed: in E.P. Thompson's words, a 'currency: not passed but spent'.[3] People clock in to work, and society is conceived and organised around the machine-like regularities of the 40-hour week and all those other institutions that adapt around this accordingly — the school timetable, regulated shopping hours and so on. Moreover, time is standardised: as late as the 1870s there were 80 different railway times in the US and France still had 14 regional times before the 1884 conference in Washington which introduced a World Standard Time and created 24 time zones.[4] Even today some countries persist in bucking the trend: many of the Maldives Islands, for example, have different time zones.

> 90% of new goods are no longer on the market two years after their launch

The clear logic of industrial time is towards homogeneity and synchronisation — a vision of the world perfectly captured in the production line of films like Charlie Chaplin's *Modern Times*. Around these continuous production processes the most successful businesses strove to synchronise life and leisure to suit production, with shift working, and a host of measures from hire purchase to macro management to stabilise demand.

As a result, in the 19th century policies came to focus on quantities of time: regulations for working time were introduced in Britain in 1802, in Prussia in 1839 and in France in 1841. Political struggles were mounted to cut the working day, guarantee holidays and sick leave, with celebrated successes like France's 1936 *congés payés*. Other policies established fixed age rights: for example, rights to schooling or pensions. For unions the key was to take control over time — and reduce it. For employers it was to get the most out of the time they had purchased.

3   E.P. Thompson, 'Time, Work-Discipline and Industrial Capitalism', *Past and Present* 38 (1967).

4   D. Howse, *Greenwich Time and the Discovery of Longitude* (Oxford: Oxford University Press, 1980); S. Kern, *The Culture of Time and Space 1880–1918* (London: Weidenfeld & Nicolson, 1983).

Post-industrial time is different again. Like the programmable digital watch, it is even further removed from nature, endlessly flexible and malleable. Activities can be precisely synchronised (to the nanosecond on modern telecommunications networks) or through techniques such as just-in-time and zero-carry-forward. And as information comes to dominate the economy, time loses its materiality. Values become less solid than in the age when buildings or steel were at the heart of the economy. Obsolescence becomes the norm, and many of the most valuable things are shaped by their half-life (the time it takes for them to lose half their value) — commodities such as chemical or genetic information, financial data or computer software. According to the philosophers, an extended present replaces the traditional distinctions between past, present and future.[5]

As knowledge takes on a greater economic significance societies invest far more in time than in things: investment in the hours needed for knowledge and learning (most of it spent on schooling) is now greater than traditional capital investment in almost every industrialised society.

In place of the ordered shared rhythms of the industrial age, time also becomes personalised and customised with the help of domestic goods (like the freezer or the VCR) that can store up services, to be drawn down at will. With personalisation also comes another important feature: just as the mechanical time of the industrial era helped to speed work up, so does the programmed time of post-industrialism. With the help of technologies, it becomes possible to perform many different tasks simultaneously, whether in work or in leisure, and to distribute functions in time and space. As Andrew Grove, chief executive of Intel, put it, 'We'll all be able to work ourselves to death — because ubiquitous computers mean that our work will

Pre-agrarian hunter-gatherers spent only 15 hours each week engaged in work

5 Barbara Adam, *Time and Social Theory* (Cambridge: Polity Press, 1990) provides an impressive overview of recent thinking.

always be with us. And our competitors will always be working too.'[6]

Nor is it only work that becomes more intensive. More valuable time encourages 'time deepening' and intensification at leisure. As Staffan Linder put it 35 years ago in his classic *The Harried Leisure Class*, the modern citizen finds himself drinking Brazilian coffee, smoking a Dutch cigar, sipping a French cognac, reading the *New York Times*, listening to a Brandenburg Concerto and entertaining his Swedish wife – all at the same time, with varying degrees of success.[7] Today we could add to that talking on a mobile phone and checking the email, while of course the Swedish wife would have traded her cocktails and sauna for a high-powered executive job.

These three successive time cultures have never been uniform. Today only a minority experience post-industrial time in much of their life (the customers of online banking, the very heavy users of the Internet), just as 50 years ago many remained insulated from industrial time. Even today, a tiny minority of the very rich still enjoy the same work- and stress-free existence as the aristocrats of the last century, while an equally small minority of 'new age' travellers and unemployed are deliberately trying to go back to a more 'natural' way of life.

> World Standard Time and the 24 time zones we still use today were introduced at an 1884 conference in Washington

Many analysts have gone badly wrong by trying to generalise minority experiences. But it is equally essential to understand that differences can be functional. Every time culture turns out to rest on its predecessors. The industrial organisation of time rested on pre-industrial norms for women who were prepared to work without contract or payment in the home. In the same way today the world of the post-industrial worker would not be viable if it did not rest on a vast array of work

> As late as the 1870s there were 80 different railway times in the US, while France still had 14 regional times

---

6 *Fortune*, 14 June 1993: 70.
7 S.B. Linder, *The Harried Leisure Class* (New York/London: Columbia University Press, 1970).

which is still essentially industrial — repetitive, mindless, regimented — as well as on a pre-industrial base of domestic labour.

Despite these differences the post-industrial model of time dominates the age. It may still be largely absent from political argument but it is centre stage in any serious discussion of business, technology, culture and the future of daily life. Its power is to offer a clearly visible goal that seems to solve many of our current problems: a 24-hour, 365-day working year that challenges not only physical or temporal boundaries but also social ones. So, in place of fixed shopping hours, it points to 24-hour shopping; instead of 8-hour days and shifts, it points to a world of infinite flexitime; in place of a fixed period of school and university education, it suggests lifelong (even just-in-time) learning; instead of a fixed period for having children, child-bearing is spread into the late 40s, the 50s and even perhaps the 60s.

Just as technologies permit instant and unbounded communication so are many seeking to pioneer new ways of accelerating things, of overcoming the limits of time. At one end there are the special programmes for gifted children designed to accelerate learning; at the other, new elixirs against ageing. The 24-hour, 365-day society has obvious advantages. It promises great efficiencies in fields such as tourism or transport by reducing the need for investment for peak loading of summer time or the morning rush hour. It promises genuine full-time operation for manufacturing, distribution and retailing with obvious benefits for productivity. But its greatest promise is political and philosophical — the achievement of genuine personal autonomy, escape from the weight of tradition, and the chance for people consciously to shape their own biography. Just as it is in the economy that the techniques of time management have developed fastest, so is it at work that this promise has first become visible. In those parts of the labour market where demand is high, employees can now make tough demands for job sharing, sabbaticals, flexible hours and parental leave.

Working time regulations were introduced in Britain in 1802, in Prussia in 1839 and in France in 1841

At first glance the labour market is indeed coming to resemble the vision of the 24-hour society. Britain, which first experienced the rigours of the industrial time culture, is now pioneering its successor. Unlike other European countries Britain has now lost for good the standard working patterns of the past, with many more people working both shorter and longer hours. Shift working has spread from the emergency services and manufacturing to many offices and computer firms, and after many false starts telework is becoming more mainstream, not only for consultants and journalists but also for secretarial and data entry workers.

Parallel changes are happening on the home–work interface. Millions of women have moved into the workforce during the same period that there has been a massive increase in domestic service work, in childcare services, and in domestic capital spending (on washing machines, microwaves and so on).

But the results have been nowhere near as benign as the optimists forecast. Even at the top of the income scale, in the jobs of the cognitive elites, this new environment has not been without costs. For them the main problem is that the nature of the value they are selling seems to entail ever greater pressures to work harder. Elite jobs demand not only innate intelligence and knowledge but also interpersonal skills, networks of contacts and trust. These skills are not easily transferable and, with these types of work, three workers working 50 hours each week are much more efficient than five workers each working 30-hour weeks, and most managers need to take work home several times a week.[8] Needless to say, these long hours impose heavy costs on families and on people's capacity to maintain a hinterland of other interests.

France introduced *congés payés*, paid leave, in 1936

At the bottom of the income range, flexibility is also more often imposed than sought — and in many countries, notably the US, real earnings at the bottom of the scale have declined just as working hours have risen. Many of these are in domestic services, cleaning

8 Andrew Jack, Makoto Rich and Emiko Terazono, 'The Daily Grind of the 7–9', *Financial Times*, 18 January 1995.

and catering, jobs that have scarcely changed over a century. Others are temporary jobs, many held by students who now need to earn more to finance their studies. And at the bottom there is a stubbornly high minority of no-earner couples (around 15% of households for the last decade), suffering persistent unemployment and non-employment.[9]

The key point is that although many of these changes are resulting in a more fluid and flexible labour market, few have yet experienced increased autonomy in any measurable way, and few feel that they have achieved a better balance between work and life.[10] Despite the rhetoric, very few workers are really in a position to customise work to their own needs.

A more flexible work environment appears, paradoxically, to have brought less personal autonomy, greater insecurity, more stress and less satisfaction. Change is coming more by imposition and fear than as part of a rising tide of freedom. 'Functional' as opposed to 'positive' flexibility seems to be the order of the day. So, although many may not fear change in itself, many do legitimately fear being on the receiving end of change, particularly those seeing their jobs swept away in new waves of automation. (As Jacques Attali put it, 'Machines are the new proletariat. The working class is being given its walking papers.'[11])

The French government introduced the statutory 35-hour week in the late 1990s

The symptoms of unease are particularly evident in the UK, which now works harder than other European countries, takes fewer

---

9   This phenomenon is analysed in Paul Gregg and Jonathon Wadsworth, *Oxford Review of Economic Policy* 11.1 (Spring 1995). A paper, 'A Short History of Labour Turnover, Job Tenure and Job Security 1975-93', presented to the International Year of the Family Conference in Autumn 1994 also showed that between the late 1970s and 1990 the probability of a workerless household having at least one working member a year later fell from around 60% to 25%.

10 NOP poll commissioned by the TUC (Trades Union Congress) and reported on in *The Pros and Cons of Part Time Working* (TUC, March 1995); TUC, 'Part Time Work', in *The Pros and Cons of Part Time Working*.

11 J. Attali, *Millennium: Winners and Losers in the Coming World Order* (New York: Random House, 1991).

holidays and suffers higher stress,[12] with time off work for stress-related illnesses up 500% since the 1950s. According to UK mental health charity Mind, job stress costs the UK up to 10% of GNP per year.[13] Some of the stress and sickness can be ascribed to the inherent problems of cultural change. Every shift in time culture has been fraught with problems. The mind-sets for industrial work took generations to establish. They were as alien to people brought up in farms and villages as post-industrial work is to people brought up with the security of a job for life and a paternalistic welfare state. The rise of industrialism saw bitter battles as rural workers were forced into the rhythms of the factory. Today, few are psychologically or culturally well equipped for a fluid work environment of contingent relationships, although, unsurprisingly, the young are far better placed to adapt than the old. Even when employers try to give employees greater flexibility over how and when they work, they meet resistance.[14]

A stubbornly high minority of no-earner couples, making up around 15% of households for the last decade, suffers persistent unemployment and non-employment

Periods of rapid change inevitably leave people disoriented. But much of the resistance to change must be ascribed to bad leadership and a failure to devise strategies to make the benefits of post-industrial time enhance autonomy and fulfilment.

In the UK, time off work for stress-related illnesses is up 500% since the 1950s

But there is also another crucial reason why change has been so unsatisfactory. For governments have failed to adapt other institutions — particularly services — to the new needs. One study at the

12 Duncan Gallie and Michael White, *Employee Commitment and the Skills Revolution* (London: Policy Studies Institute, 1993).11 Duncan Gallie and Michael White, *Employee Commitment and the Skills Revolution* (London: Policy Studies Institute, 1993).

13 Mind, *Stress and Mental Health in the Workplace* (London: Mind, 2005).

14 'Working Hours and the Organisation of Working Time: Collective Regulation or Individual Choice?', in ETUC (European Trades Union Congress), *Time for Working, Time for Living* (ETUC Conference, 8 December 1994).

end of the 1980s found that, between 7 am and 7 pm on a weekday, the average European consumer had only a 10% chance of finding a service open at the times when they are free. Women's changing position is at the core of this issue. Services were designed around the assumption that women would be able to shop, clean, cook, collect and care between 9 am and 5 pm while men were at work.

For example, in 1939, 31% of workers went home for lunch; most could safely assume that a woman would be there. Women were effectively informal service workers taking children to doctors and schools, or caring for the sick or elderly. Their 'free time' served as a buffer in the management of time. But today with dual-earning in a majority of households, and with a rising proportion of single households, many of these assumptions fall away. The synchronised world of public institutions and private services is now at odds with the labour market, and the pressure is growing for a more general policy on time that acknowledges the systemic nature of time — the ways in which all the elements are interdependent. Take, for example, public amenities. The combination of fear of crime, unfriendly public spaces and badly planned transport systems means that in Britain adults now spend 900 million hours each year escorting children to school (children now rarely travel to school on their own: 38% are now taken by car, compared to 22% 15 years ago).[15]

The UK charity Mind estimates that job stress costs the UK up to 10% of GNP every year

The average journey to work has risen roughly a mile each decade since the 1950s and more time is now spent in getting to shops, schools and entertainment. Travel to work adds an average of five weeks a year to a UK employee's working life. Or take services as a whole. There have been some steps to open the service economy to fit the needs of dual-earning couples: liberalising Sunday shopping, deregulating the evening economy, and ensuring longer opening hours for schools are all examples. There have also been experi-

15   Mayer Hillman, John Adams and John Whitelegg, *One False Move* (London: Policy Studies Institute, 1991).

ments with shifting the time economy of cities,[16] some of which are described by Franco Bianchini. And recent years have begun a revolution in public services which has brought not only 24-hour access to NHS Direct but also far more weekend and evening opening, taking the UK well ahead of other European societies — often over resistance from trades unions and churches. The advent of postal and online voting is another example of simple changes that can be made to align public institutions with the new patterns.

These are some of the problems that arise from the clash between an old and a new order of time. But systems theorists add a more subtle step to the argument, one that is particularly relevant to the likely politics of post-industrial time. In more complex systems, they suggest, there is bound to be more enforced waiting because so many more things need to be co-ordinated.[17] In parallel, they argue, waiting is bound to become more visible in more prosperous societies, because it is perceived

> A study at the end of the 1980s found that, between 7 am and 7 pm on a weekday, the average European consumer had only a 10% chance of finding a service open at the times when they were free

in relation to the many more things we would like to be able to accomplish. So, although some imaginative ways of using dead waiting time are being experimented with — interactive terminals at railway stations and bus stops, for instance — time planning and time reduction strategies are almost certain to rise up the agenda.

For many engaged in this argument there is a simple solution: to cut working hours. Over the last 150 years annual working time in the industrialised countries has fallen steadily: from around 3,000 hours to between 1,400 and 1,800 hours (Japan still stands at over 2,000 hours, and worked 2,400 hours as recently as 1960). The main reason for this fall of course is productivity: one hour in 1990 is five times as productive as it was in 1830.[18] But in the 1980s the decline

16 See examples set out in Charles Landry and Franco Bianchini, *The Creative City* (London: Demos, 1995).
17 Niklas Luhmann, *Politische Planung: Aufsatze zur Sociologie von Politik under Verwaltung* (Opladen: Westdeutscher Verlag, 1971).
18 ETUC, *op. cit.*

in working hours stopped. The average British working week is now 43.4 hours, an hour longer than in 1983, and the decline has also ceased in other countries.

Although economists have always been sceptical about work sharing, during the 1980s a variety of schemes involving unions, employers and governments in mainland Europe were put into practice, mainly inspired by the need to stop redundancies. The French government led the way in the early 1980s by encouraging firms and unions to agree 'solidarity contracts' (involving early retirement, wage restraint, new jobs for the young and different working time flexibility options) and backed this up with a decree reducing the length of the statutory working week by one hour to 39 hours a week, before introducing the statutory 35-hour week in the late 1990s.[19]

In 1939, 31% of workers went home for lunch

In Germany, by contrast, the main pressure has come within industry, where IG Metall negotiated a series of deals for shorter working weeks and greater flexibility to protect jobs. The most celebrated was struck with Volkswagen and involved a four-day week and a significant, but not proportionate, cut in pay (16% instead of 20%). A total of 20,000 jobs were protected while another 10,000 were saved by encouraging initial part-time work for the young, early retirement and training sabbaticals.[20] Similar deals have been struck between unions and both Bell and Chrysler in Canada.[21] In each case high

Adults in Britain spend 900 million hours each year escorting children to school.

19 The law meant that, if employers cut working hours by 15%, accompanied by a pay reduction, and recruited within six months new employees equivalent to at least 10% of the annual average workforce, their social security contributions would be cut by 40% in the first year and 30% in the next. ETUC, 'Working Hours and the Organisation of Working Time: Collective Regulation or Individual Choice?', in ETUC, *op. cit.*

20 Giuseppe Fajertag, 'Working Less to Enable Everyone to Work?', in ETUC, *op. cit.* Both give detailed explanations of the VW model.

21 Full details of these and similar schemes can be found in the *Report of the Advisory Group on Working Time and the Distribution of Work*, published by the Canadian government in 1994.

productivity, powerful unions and a strong competitive position made quite generous deals possible. Even Japan — which still has the longest annual hours at 2,000 — has made much of the need to cut working hours. A government report in 1989 stated 'we must work hard but as a result we must take a rest'. Few concrete policies have been implemented but a cultural shift in attitudes to work may be achieving the same ends as a 'grasshopper generation' (*kiri-girisu*) refuses to work overtime and nights, organising life around the three Vs: villas, visits and visas. Their commitment to leisure represents an overt rejection of the overwork culture that causes 10,000 deaths each year from *karoshi*,[22] and that, according to one survey, means that 124,000 of Toyota's 200,000 workers suffer from chronic fatigue.

So far, however, few of these schemes have achieved much success. Some deals at the firm level fudged the issue of costs, and thus potentially jeopardised jobs in the long run. Others — like VW — make it hard to generalise because they involved relatively highly paid workers who could afford modest cuts in pay.[23] Meanwhile, government policies have been either too costly or too inflexible to achieve their desired results. France continues to be plagued by particularly high unemployment (and ironically has seen less redistribution of work from full- to part-time and from men to women than deregu-lated Britain). There is little strong positive evidence from the 35-hour week experience. Schemes to encourage early retirement so as to free up jobs for the young have proven prohibitively expensive (mainly because of the added pensions costs) as are policies to directly subsidise work sharing, such as Canada's Work Sharing programme, not least because they subsidise many workers who would otherwise get new jobs. So it would be wrong to place too much faith in such schemes. Many feel like hangovers from a now

Children rarely travel to school on their own: 38% are now taken by car, compared to 22% 15 years ago

22 Death from heart disease or stroke caused by overwork.
23 Even after the introduction of a four-day week, the average VW worker still earns more than the average full-time German worker.

lost era when men worked a standard working day which could be legislated downwards. Others (like the EU directive and Japan's recent legislation) have so many exemptions that they are virtually meaningless. In general, micro solutions offering

The average journey to work has risen by roughly a mile each decade since the 1950s

individuals a range of choices have worked better than macro solutions that have tried to legislate time. One of the reasons may be that top-down rules clash with the principle of autonomy.

Although a significant number would welcome shorter hours, others want to work hard, either because they need the money or because they enjoy it. Even on high incomes many choose to work harder because they value the extra money – and the opportunities it gives for more intensive leisure activities – more than the extra time. For them, it's better to work to buy a CD or a tropical holiday than to create time to potter around the garden.

If macro solutions have proven unsuccessful, what could be done to better achieve this balance? Clearly for many the priorities are fairly basic: to have the marketable skills to earn

Over the last 150 years annual working time in industrialised countries has fallen steadily from around 3,000 hours to between 1,400 and 1,800 hours. Japan still stands at over 2,000 hours, and worked 2,400 hours as recently as 1960. The decline in working hours stopped in the 1980s

more and bargain for better treatment. But even for them, life would be easier if it were possible to trade money and time in more flexible ways. Hence the importance of rights to ask for more flexible working arrangements, rights for part-timers and rights to leave. Hence too the importance of rethinking the time patterns of public and private services.

However, on its own this agenda of autonomy and flexibility doesn't go far enough. Autonomy is necessary and desirable, but few would want to live in a totally flexible, unbounded time economy. Nor is such an economy likely. Throughout history times of transition have led some to extrapolate from short-term trends to forecast permanent chaos, anarchy or fluidity. But the lesson of human history is that periods of flux tend to be followed by the creation of new fixed structures and reference points which in practice make flexibility easier to organise. Just as in language or music fixed rules and grammar enable an almost infinite flexibility, so in life does variety rest on some things being fixed. In the medium term two main barriers seem likely to block the achievement of the totally flexible world sometimes implicit in descriptions of post-industrial society. The first barrier concerns basic human needs. We now know much

> The average British working week is now 43.4 hours, an hour longer than in 1983

more about how time is inscribed in our biology. Psychologists have learned how to get over our 25-hour clocks, our circadian rhythms, and fit these to 24-hour factories or international air travel. But in other respects we may be coming to learn about the limits to flexibility. Parenting is one obvious example where the clash between biological instincts and the demands of society or the workplace is leading to huge stress and may well encourage a swing back to the virtues of encouraging more relaxed enjoyment of the parenting experience for both men and women. More generally we are learning that few are well suited to chronically uncertain work environments. The second barrier is interdependence. Individual choices inevitably impinge on others. This is less true for the 23-year-old com-

> In Japan, the overwork culture causes 10,000 deaths each year from *karoshi*, which literally translates as 'death from overwork'

puter programmer than it is for someone with children. Too much personalisation of time may have unwanted effects. One obvious example is the way that just-in-time production methods ironically increase the number of journeys by trucks, worsen congestion and thus eat into the free time of others. Another is the way that the failure to preserve safe streets forces parents to drive their children to school, and the resulting congestion eats into free time for others.

These limits suggest that, as well as flexibility, we also need new reliabilities (what Christopher Freeman and Luc Soete describe as a new 'vertebrate structure'[24]) to replace those of the industrial age

---

24 Chris Freeman and Luc Soete, *Work for All or Mass Unemployment? Computerised Technical Change in the Twenty-first Century* (London: Pinter, 1994).

— reliable rhythms around which people can cope with a far more fluid world of time. If recent years have seen a marked de-institutionalisation of work, with the decline of trades unions and managerial hierarchies, the challenge of the next period may be to achieve an effective re-institutionalisation. The fully flexible model sometimes implies no structures between the individual and the task, and no sense of the workplace as a source of community, identity and psychic fulfilment. But for work to be more than just a source of income it must involve some reliable elements. What might these be? Conventional wisdom assumes the end of the job, of traditional employment and thus of security. The problem with this model is that it leads to irrational effects: under-investment in human capital, uncommitted workstyles, and cultures of distrust. Some employers understand this and in parts of the service sector and hi-tech manufacturing firms are experimenting with new forms of commitment, involving job security and investment in skills in exchange for greater task flexibility. But this model is likely to be relevant only for a minority. For others an alternative model may be the development of 'deployers' — firms with a long-term relationship with individuals, which deploy their labour to others. This model already exists in many fields: in clerical and secretarial work, and increasingly higher up the income scale. A good example is the Cooperative Home Care Associates organisation based in South Bronx (whose achievements inspired the formation of a national Cooperative Healthcare Network), a co-operative of hundreds of women who have been taken off welfare to work in home healthcare, financed in part through the benefits savings to the state. This could turn out to be one future for trades unions, as intermediaries between the individual and the labour market. These types of arrangement could provide a more reliable core relationship in the workplace.

124,000 of Toyota's 200,000 Japanese workers suffer from chronic fatigue

78% of people aged 25–34 years would continue to work even if there was no financial need (compared to 66% of those aged 45–54)

I have already hinted at the anxiety lying behind much of the debate over time: the sense that material progress has not brought much improvement in how we use our time. It is an anxiety confirmed by every study of time use, showing that most work and leisure is not used well, whether the goal is happiness or personal development.[25] The evidence we have shows that the most fulfilling activities are those entailing autonomy, demands on skills, and absorption. These have been described by the psychologist Mihaly Csikszentmihalyi as the characteristics of flow. Flow can be found in many kinds of work — sports, music, arts and crafts — indeed any kind of task with structured goals that places a demand on people's skills.[26] It goes without saying that much work has been short of flow and it is this legacy of work as a burden that explains the attractions of many for greater free time.

But why should we assume that leisure is inherently superior to work? Sociological analysis of time has long acknowledged that it is the quality of time experience that is key. And the analysis of flow tells a very different story to that casually suggested by advocates of a liberation from work. It shows that many people — and not just professionals — find leisure less fulfilling and challenging than work. In one major survey, for example, even assembly line workers experienced 'flow' more than twice as often in work as in leisure. The reason is simple: in general, work is more likely to lead to flow, because of its structured tasks, feedback and challenge, whereas far too much leisure is passive, shapeless, unchallenging and literally unrewarding. This is surely one reason why 78% of people aged 25-34 years would continue to work even if there was no financial need (compared to 66% of those aged 45-54). It suggests that Ronald Reagan's famous comment, 'They say hard work never did anyone any harm, but why risk it?', was rather off the mark.

> On average Britons spend over 25 hours each week watching television

Indeed, although greater leisure has led to some more imaginative uses of time, most still use it in low-intensity activities. On average Britons spend over 25 hours each week watching television.

25   Michael Argyle, *The Psychology of Happiness* (London: Routledge, 1994).
26   M. Csikszentmihalyi, *Flow: The Psychology of Optimal Experience* (New York: Harper & Row, 1990).

In the 1980s only one in four men and one in ten women had taken part in an outdoor sport or activity in the previous four weeks. Few know how to achieve the inner discipline to shape time and use leisure for genuine 're-creation', and, as Csikszentmihalyi writes, we waste each year 'the equivalent of millions of years of human consciousness [as] the energy that could be used to focus on complex goals, to provide for enjoyable growth, is squandered'.

There are some signs of a less vegetative mind-set among the young, brought up with more interactive games, able to use the remote control to multi-task work and leisure, unlike their parents and more demanding of variety, interaction and engagement. The phenomenon of 'time deepening' means that many are trying to do several things at once, accelerating leisure by driving through zoos rather than walking, or substituting intensive sports like aerobics for tennis or golf. But most people are still bad at managing their time, preferring for the most part activities which fill time rather than use it.

Today the momentum of post-industrialism is towards flexibility and autonomy. The benefits are flowing to those societies and individuals who can meet Italo Calvino's definition of the qualities for the next millennium: lightness, quickness, exactitude, multiplicity.[27] But these are not enough. A truly fluid and disordered world will leave the majority miserable. So alongside speed and flexibility we also need to remember the importance of balance: of mechanisms for finding useful activity for those left out by change; of public spaces for quiet and reflection like parks and churches where time stands still; of home life as well as work life. Indeed, in our view, this balance is the ultimate promise of the post-industrial order: not a subordination of all life to the requirements of a new techno-economic paradigm, but rather a more sustainable balance between the diversity of human needs.

In the 1980s only one in four men and one in ten women had taken part in an outdoor sport or activity in the previous four weeks

27   I. Calvino, *Six Memos for the Next Millennium* (London: Vintage, 1996).

People have always told others how to live and spend their time. I like Samuel Johnson's comment, 'I have all my life long been lying in bed until noon yet I will tell all young men and tell them with great sincerity that nobody who does not rise early will ever do any good.' And the less ironic comment of a vicar, author of *Friendly Advice to the Poor*, who wrote that 'the necessity of early rising would reduce the poor to the necessity of going to bed betime and thereby prevent the danger of Midnight revels'. Today we can do without such advice. We have the means to get closer to controlling our own time, our own lives. And, though, as in the phases of an earlier industrial revolution it at first makes life a servant of work, it offers the opportunity of work becoming part of life.

6

Time and money

Will Hutton and Alexandra Jones

Dost thou love life? Then do not squander time, for that is
the stuff life is made of.

Benjamin Franklin

Time is the 21st-century must-have. Everyone seems to want more.
For many of us, more time would mean we could feel in control; we
could spend time with family and friends, help in the community,
or just relax and enjoy ourselves. More time would mean being able
to make that tight deadline at work, to make some kind of dent on
the growing mountain of 'things to do'. And the market is doing its
best to oblige, offering a range of panaceas, from microwave meals
to the booming business in time management books, the better to
help us manage our hectic lives.

Time, after all, is money. Time spent doing one thing is potentially 'wasted', time that could have been spent doing something
else, something better, something more productive. The idea of the
leisure-rich society has retreated in favour of a busy society. If
Americans had maintained their productivity levels after World

War II, they could take every other year off; but, instead of using technological advances to decrease our workload, we try to cram more in.

For most of us in the 21st century, our understanding of time is heavily embedded in capitalism. It draws on an original, simple idea that 'time is money', a currency that we can spend or save, a resource, an economic variable. But how has this attitude come about? How does it affect the way in which we value and measure time? What does it mean for the way we live our lives? And how is it changing?

Time is the 21st-century must-have

A brief history of time and capitalism

Time and capitalism are old adversaries. Since the Industrial Revolution, when factories created workplaces outside the household, 'time at work' has been seen as a commodity to be regulated — and as such it moved to the forefront of the battle between labour and capital.[1] The idea of productivity — how much product someone could make in a fixed period of time — quickly became a key measure of industrial capitalism's efficiency. Adam Smith famously gave the example of pin manufacture, arguing that one person making pins might not even manage one per day, but that ten people, with division of labour, might make 48,000, increasing productivity to 4,800 pins per worker.

Long working days and inflexible working time were identified by Marx as being in the nature of capitalism, with market forces keeping wages low and so forcing individuals to work longer hours

---

1 Engels argues that it was when work was moved outside the household that 'paid work' and 'household work' became separated, men gained ownership of private property and women came to be dominated by men. See F. Engels, *The Origin of the Family, Private Property, and the State* (Harmondsworth, UK: Penguin, 1986 [1884]).

in order to earn a subsistence.[2] This was clearly to the advantage of employers: in manufacturing, only a certain number of products could be produced for each unit of time, therefore long working days became a means of increasing productivity and creating surplus value. Work schedules became increasingly inflexible as employers obsessed over maximising the productivity of their machinery and their employees, enabling them to regulate and measure production.[3] Time equalled money.

Unions fought long and hard over working time: for health and safety regulations; for a weekend and an 8-hour day. The slogan in America was 'Eight hours for work, eight hours for rest, eight hours for what we will'.[4] Successes were slow to arrive. The earliest time legislation in the UK, passed in 1802 to limit the hours children could work in factories and cotton mills, proved difficult to enforce and was largely ineffective. It took seven Factory Acts in the 19th century to ensure that adults and children were offered reasonable working conditions.[5] However, by the mid-20th century, a 40-hour week, 8 hours a day, had become roughly the norm.

It took seven Factory Acts in the 19th century to ensure that adults and children were offered reasonable working conditions

These working time arrangements, hard fought and hard won, were embedded in the dominant capitalist and social system of the day. Dr Steffen Lehndorff of Germany's Institute for Work and Technology argues that the 8-hour day synchronised patterns of time use throughout society, and created the original space for the

---

2 P. Meikins, 'Confronting the Time Bind: Work, Family and Capitalism', *Monthly Review* 49 (February 1998): 1-13.

3 E.P. Thompson, 'Time, Work Discipline and Industrial Capitalism', in *Customs in Common* (New York: The New Press, 1991): 352-403, cited by Meikins, *op. cit.*.

4 Used by American trades unions.

5 Including giving women the right to four weeks off work after giving birth, obliging employers to ensure their machinery was (relatively) safe and raising the minimum age for child workers to 11 years (up from 10 years).

development of a consumer society.[6] This consumer society increased the capacity of the market to absorb standardised, mass-produced goods and formed the bedrock of post-World War II growth.

The 8-hour day was built on the model of the dominant male breadwinner. This was the 'two for one' deal, as Shirley Burggraf[7] put it, whereby for every employee — usually a man — an employer would benefit from the labour of one person at home, usually a wife, for free. Until the early 1960s women were expected to leave work on marriage, the idea being that wives would dedicate their (unpaid) time to childcare, housework and community responsibilities, freeing up their husbands to concentrate on paid work. Even 'time brokers' in paid work — PAs — were predominantly (single) women, perhaps reflecting the view that a woman's job was to support a man's work. This paradigm was so deeply ingrained that it provided the basis for the entire post-World War II social security system, constructed around the assumption that men would be working 40-hour weeks for 40 years, with women able to rely on their husbands for support throughout their lives. The 21st-century reality is clearly quite different — and so are the ways in which we need to measure and value time.

Demanding capitalism

The labour market has changed dramatically over the past 50 years. Globalisation has made competition more international while intensifying the importance of truly local, customised services for

6 S. Lehndorff, ' "Tertiarisation", Work Organisation and Working Time Regulation', paper presented at the international conference: *The Economics and Socio-economics of Services: International Perspectives* (Lille/Roubaix, 22–23 June 2000).

7 S. Burggraf, *The Feminine Economy and Economic Man: Reviving the Role of Family in the Post-industrial Age* (Reading, MA: Perseus Books, 1997).

increasingly demanding customers.[8] Large suppliers compete directly with one another, while at the same time long-term average growth rates have declined. Technology permits far greater reach to customers, at the same time arguably facilitating greater demands for access and customisation. This means that time is becoming an ever more important factor in competition, not only in providing goods and services quickly, as and when required, in increasingly tight markets, but also (more importantly) in opening up new markets.[9]

Financial markets place a growing emphasis on short-term performance, not year on year but minute by minute, hour by hour. It has even been argued that the goods-based economy is becoming a function of the financial economy. The result? Demands for high financial returns in the short term. Investment and careful building-up of market share are sacrificed to boost profitability in the immediate future. It is the share price now that matters, not the higher return that could be achieved in five years' time. In fact, the very structure of the financial economy prioritises short-term gain.

The operation of the capital system relies on liquidity, the ability to reverse a lending or investment decision and return to holding cash. This is the premise on which the City operates, as a series of related markets in which financial assets of any type are readily tradable for cash; and this emphasis on liquidity, which grows as the influence of financial markets grows, creates incentives to be risk-averse in investment strategies, to invest in short-term returns rather than longer-term growth. Shareholders can 'exit' rapidly, and so companies work to secure their loyalty through high and growing dividends. The result is financial engineering rather than business building.

> In reality the market lends more importance than is justified to the short-term future, ignoring the long-term damage this can inflict

Supporters of the system argue that the freedom to buy and sell at will ensures that companies operate to maximum efficiency, with

8   A. Rajan, B. Martin and J. Latham, *Harnessing Workforce Diversity to Raise the Bottom Line* (London Human Resource Group, 2003).

9   Lehndorff, *op. cit.*.

poor financial returns punished through firing someone or through takeovers. Yet this assumes both that the market accurately values future returns and that there are no efficiency losses from the constant re-evaluation of a company's worth.

In reality the market lends more importance than is justified to the short-term future, ignoring the long-term damage this can inflict, particularly on relationships with workers, suppliers and customers. As Harvard business guru Professor Michael Porter argues:

> The operation of the entire capital system . . . creates a divergence of interests among shareholders, corporations and their managers that impedes the flow of capital to those corporate investments that offer the greatest pay-offs.[10]

Markets operate on the assumption that they can accurately value returns now and in the future but, in reality, they prefer now to later. This reflects a well-known psychological phenomenon that shows that we are wholly inconsistent in the way we rank rewards over time. Offer an individual £100 now or £150 in six months, and he or she will almost always take the £100 now; we prefer rewards in the present, while rewards in the future have to be much higher than they should rationally be in order to persuade individuals to accept them.

This 'time-discounting' feature affects the stock market dramatically. A company with a strategy to do well now and excellently in three years' time could be taken over, with shareholder approval, by a company that will do very well now, but less well in the future, resulting in an overall loss for the shareholder — but an immediate gain. So the 'time is money' principle is too simplistic; in reality, money now is better than money tomorrow. This short-termism damages companies by rating immediate profits more highly than larger, deferred profits, and by destabilising what could be very good, sustainable, visionary organisations. Making current shareholder

Offer an individual £100 now or £150 in six months, and he or she will almost always take the £100 now

10 M. Porter, 'Capital Disadvantage: American's Failing Capital Investment System', *Harvard Business Review*, September/October 1992.

value and short-term corporate profitability the sole yardstick of measurement distorts performance.

In one respect consumers are served well by this system, since their ability to 'shift allegiance with the click of a mouse'[11] means that companies have to run to keep up. But this also encourages continuous cost cutting, reinvention and drastic measures to produce immediate results, which in turn translates as short-term thinking among managers.

These incentives run counter to Jerry Porras's and James Collins's findings in their six-year study of what makes a great enduring company. In *Built to Last*[12] they found that visionary companies were those that had built organisational systems lasting beyond the tenure of one individual and had a purpose beyond profit – such as Walt Disney's core value of 'Making people happy'. And they also argued that visionary companies needed not only to be resilient to change but to actually have a passion for it, which in practice meant taking risks, such as Boeing 'betting' the company on the 747. This, the authors recognise, flies in the face of established financial wisdom. The market's conception of an acceptable rate of return is at odds with a strategy to take any time to build up a company, or to change tactics. The focus is the next quarter's results, not on enduring success. CEOs, whose tenure is diminishing, are being rewarded for immediate successes, and encouraged in the belief that their successor can pick up the pieces: an attitude that runs throughout organisations, where managers are encouraged to 'make their mark' and move on. This reinforces the belief that, in the financial market, what matters is not yesterday, not tomorrow, but now.

> Visionary companies needed not only to be resilient to change but to actually have a passion for it

Research conducted by The Work Foundation in the Work and Enterprise Panel of Inquiry examined the factors that contribute to

11   R.B. Reich, *The Future of Success* (New York: Alfred A. Knopf, 2001).
12   J. Collins and J. Porras, *Built to Last: Successful Habits of Visionary Companies* (New York: HarperBusiness, 2002).

high organisational performance.[13] As a result, we constructed the
HPI (High Performance Index), an analytical framework that en-
compasses the whole of an organisation's strategic agenda, ranging
from shareholder value to customer focus, innovation performance
and human resources, underpinned by a set of leading performance
indicators. The results of our research are startling: businesses that
perform significantly better manage all the capitals of the HPI and,
critically, the trade-offs between them.

In other words, it is crucial to make the right judgements between
customer, employee and shareholder interests. Focusing on one
area disproportionately and not working through connections with
other areas of the HPI is likely to damage performance. So concen-
trating solely on shareholder value means strategies could fail to
connect with the organisation and would not ultimately deliver.
But those companies that do make the connections and trade-offs
can deliver stunning value growth. Organisations in the upper quar-
tile of the HPI are, on average, 47% more productive than those at
the lower end. Yet to make these trade-offs requires the investment
of money and time which the markets mitigate against. Narrowly
focused short-termism is not the way to deliver value in the longer
term — and yet it is what we continue to value in the business world.

Some have argued that this increasing emphasis on 'now', rather
than later, is a product of an increasingly consumer-focused society
— and is having detrimental impacts on health. The economist John
Komlos and colleagues suggest that 'recent trends in obesity are
related to an increase in the marginal rate of time preference'.[14] He

13 The research involved interviewing 20 chief executives of FTSE 250 com-
    panies, conducting case studies in 30 private-sector organisations, holding
    select committee-style evidence sessions and quantitatively analysing
    existing survey data. From this we constructed a High Performance Index
    and, then, employing a representative sample of 1,000 companies, we re-
    searched and modelled the correlation between company performance, as
    represented by factors such as profitability, growth and sales per employee
    and the HPI and its leading indicators.

14 J. Komlos, Patrick J. Smith and Barry Boggin, 'The Rate of Time Preference
    and Obesity: Is There a Connection?', *Discussion Papers in Economics* 60
    (Munich, 2003).

argues that the higher the rate of time preference, the more likely it is that individuals will discount the future health risks associated with their current eating habits and lack of exercise. If this is indeed the case, it has implications for many other activities with long-term negative effects, such as smoking, getting into debt, or polluting. The growing emphasis on now, regardless of consequences, will have damaging effects on far more than just business.

21st-century workers

The financial system does not stand alone, however. It is informed and reinforced by the social and institutional framework that has grown around it. While competition was less rampant and the male breadwinner model more dominant, society endorsed the idea of men working an 8-hour day through everything from cultural norms and stereotyped advertising, to old boys' networks and all-male clubs. With the financial system shifting in response to intensified competition, improved technology and a more global market, there have also been shifts in the social norms that held up the status quo.

The growing emphasis on now, regardless of consequences, will have damaging effects on far more than just business

A key change is in the nature of the psychological contract. Robert Reich argues in *The Future of Success* that individuals are much less certain now about 'tomorrow'. The pace of change has left them unsure about future earnings, which organisation they will work for, or even what they will be doing.[15] He argues that a demand is being created for workers who respond to the uncertainties, cost cutting and the intense nature of competition by working longer and more intensively in paid work. Technology has also facilitated long working hours by ensuring that individuals can always be on call — at home, at play, abroad; laptops and mobiles

15   Reich, *op. cit.*.

allow 24/7 accessibility. It is thus individual workers, their families and friends who have become the most time-compromised.

Despite the fact that over the last ten years job tenure has increased, on average, from six years two months to seven years four months,[16] the concept of a job for life is dwindling and the idea of personal employability growing. Individuals see themselves as responsible for developing their own skills and careers, recognising the reality that, in many companies, regular cost cutting and changing product needs mean they face greater competition to keep their job. This can reinforce employees' willingness to work long hours in each job they have, and many individuals thrive in this intense way of working. But the changes in the labour market mean that tensions are arising between the way in which the financial system encourages organisations and individuals to work, and our social framework.

The labour market is changing dramatically. The two-for-one offer is, for the most part, over: 46% of the UK labour force is female, and 80% of the workforce growth to 2010 will be among women. While women have always worked, the sheer scale of their labour market participation has undermined the dominance of the male breadwinner model. Eight-hour days and five-day weeks may have worked well when men could rely on unpaid care at home to look after everything outside paid work, but now many households cannot afford or do not wish to function this way. And, with dual-income households on the rise alongside an increasing demand for time spent in paid work, a serious tension is emerging between time in paid work and time outside it.

> While women have always worked, the sheer scale of their labour market participation has undermined the dominance of the male breadwinner model

As work and the rest of life are more and more integrated, supported by technology, the rest of life is subject to a squeeze. The rise of 'work–life balance' in the political and organisational agenda and the concept of 'family-friendly' working are just two of the most

16 R. Taylor, *Britain's World of Work: Myths and Realities* (ESRC Future of Work Programme, 2001).

obvious expressions of the conflict between work time and everything else.

Despite these changes, the way in which we work has not changed
a great deal. We continue to regard time as an economic variable,
like capital or labour, that can be made more effective through
maximising the amount done in the shortest interval possible. But
this assumes that there is a linear relationship between time and
productivity, that simply does not exist. For example, individuals
get tired, and often start making mistakes after a
certain length of time working. The Chernobyl
disaster is just one example of an accident where
fatigue caused by overwork was to blame. And, although many would argue that time is a helpful
proxy in manufacturing, Henry Ford disproved
this in the 1920s. Having experimented with vari-
ous different work schedules in the 19th century, he finally decided
to introduce a five-day, 40-hour week for six days' pay in 1926. This
was because he had found that workers in his factories could
produce more in five days than they could in six days; more time,
he had found, did not necessarily equal more money for him.[17]

> Henry Ford found that workers in his factories could produce more in five days than they could in six days

Now the economy is dominated by the service sector, where it is
even more difficult to predict productivity per hour; yet we continue to use time as a proxy for measuring productivity because we
find it so difficult to define the outputs and outcomes we want.
Time is seen as a variable that can be flexed when the going gets
tough. Even organisations that do focus on outputs, for instance
through projects, have a tendency to do this. In fact Richard Scase,
professor of organisational behaviour at the University of Kent,
argues that, when the principles of decentralised project management are applied under an Anglo-American model, a culture of long
hours results: resources are inevitably underestimated when work
processes are broken down into project-driven targets.[18]

17 H. Ford, 'Why I Favour Five Days' Work with Six Days' Pay', *World's Work*,
    October 1926: 613-16.
18 R. Scase, *Living in the Corporate Zoo* (London: Capstone Publishing, 2002).

Many companies believe that employees should do in their own time things that do not directly generate money, such as administration or answering emails. A study of an accountancy firm by Suzan Lewis, professor of organisational and work-life psychology at the University of Manchester, found that accountants were regularly encouraged to underestimate the time it would take to complete a project to make a bid more competitive. 'Invisible hours' of work are then needed to complete the project according to the deadline and accountants collude in this in order to succeed. The employees themselves acknowledge that in order to 'get on' in the organisation they need to meet these deadlines — which requires long hours working, often in the office, since 'face time' is valued as a sign of commitment.

Which means that, for individuals with responsibilities or aspirations outside work, it is incredibly tough to succeed. The difficulty of combining paid work with caring responsibilities, for example, is demonstrated starkly in any number of statistics, such as the cost in lost earnings of being a mother of two with medium skills over her lifetime — £140,000[19] — or the fact that senior female executives are highly unlikely to have children, in contrast to senior male executives. The pay gap, and the fact that caring professions, dominated by women, are undervalued financially, demonstrate that the way the labour market currently operates is failing to enable men and women to combine paid work with other things in their lives.

> Accountants were regularly encouraged to underestimate the time it would take to complete a project to make a bid more competitive

And, despite the fact that they occur in many countries, the tensions between paid work and the rest of life are seen as being something that individuals need to manage. It is individuals who are increasingly making choices between 'time' and 'money' — time with friends and family or being paid for time with their company. The working time arrangements we once relied on are fast becoming out of date.

19 K. Rake (ed.), *Women's Incomes over the Lifetime* (London: UK Government Cabinet Office, 2000).

Valuing time differently

The value we place on time is changing — and with it our relationship to it. Which raises questions about whether our idea about time is correct after all. The idea that 'time is

For individuals with responsibilities or aspirations outside work, it is incredibly tough to succeed

money' — that time is an economic variable like labour or capital, that time is a currency that can be bought, spent or saved — seems to be spiralling out of control and at the same time becoming dangerously irrelevant to the way we live our lives. We propose three ways in which this might be effectively addressed.

1 Time in the abstract

At the core of the tensions we are experiencing in trying to manage paid work along with the rest of life is our contradictory attitude to time. We treat it both as a currency, something that can be controlled and measured, and also as something much more complex which varies from hour to hour, day to day, which we struggle to control or measure at all.

We talk about time as a currency — we spend it, save it, buy it, sell it. Work is structured around the assumption of a tight link between time and money; people are paid by the hour, week or month; profit is calculated with reference to quarters and years. Yet all of this is possible only if you have an idea of, as Barbara Adam calls it, 'empty time'.[20] This assumes that time can be separated from content and context; it is always an abstract, standardised unit. The only way that we can value each hour as being the same is by assuming that each has the same, universally applicable value. An hour spent driving a cab is the same as an hour spent driving in a Formula 1 race: after all, both individuals are doing the same thing for the same abstract amount of time. However, while one is

20   B. Adam, *Timewatch: Social Analysis of Time* (Cambridge: Polity Press, 1995).

likely to earn £20 net, the other may earn up to £200,000: the monetary value of that time is very different. Looked at from a personal perspective, an hour spent counting paperclips and an hour spent at a friend's birthday party are, in abstract, the same length of time. In experience, counting the paperclips will create a seemingly interminable hour of boredom, while the celebrations will mean the hour flashes past, is enjoyed and is something to remember in the future.

Luhmann argues that there are four characteristics of what he calls this 'mathematical, universalised' time:

1. Independence from context and content

2. Chronological dating

3. Translatability — the capacity to compare non-simultaneous stretches of time

4. Mental reversibility, by which we are able to repeat what is fundamentally irreversible[21]

In other words, we treat time as a commodity. Efficiency means producing or performing something in the shortest possible time. Being profitable means spending as little money as possible on labour time. Being competitive means being faster than your rival. These are all values that are increasingly reflected in our changing business world, as we use the financial markets as the benchmark for judging performance.

Yet this is not the reality of our experience of time. As individuals we recognise that not all time is the same, neither in monetary value nor in personal value. We can earn more for some activities than for others. We all have hours that pass more quickly because we are enjoying ourselves, and hours that pass more slowly because we are bored. Einstein famously said, 'Put your hand on a hot stove for a minute, and it seems like an hour. Sit with a pretty girl for an hour, and it seems like a minute.' Time used for different activities and at different moments has a different value and meaning. What about

21 Quoted in Adam, *op. cit.*

time that falls outside employment relations? How long should it take to get to know someone? Can a relationship be made more efficient? Speed-dating suggests that at least the initial stages of relationships can. But this ignores the fact that there is no sense of 'enough' time spent with children, for instance. And the quality of our time, the way in which we interact, always changes. Which hour is more productive than another? Or is that the question we should really be asking?

Time does not exist out of context — and so pretending that it does is futile

So, while we tend to regard time as something that can be measured in the abstract, it is experienced differently depending on the individual, the circumstances, whether it passed today or yesterday or whether it will pass tomorrow. Time does not exist out of context — and so pretending that it does, and that we can effectively plan ahead or manage it, is, in many ways, futile.

2 Success = market success?

Barbara Adam has written extensively on the relationship between time and society and argues that we increasingly measure time as a whole in the same way that we measure time spent in employment. Market values of now rather than later, of speed and of demonstrably productive use of time may become the filters through which we value all our time. Success becomes equated with 'busyness', and time may be seen as 'wisely spent' if it is invested in building up career skills and contacts — even, ironically, if the payback is not now but later, as it is a measurable return on one's investment. Time invested in caring for others, relaxing or working in the community, which often has a much larger return later in life through health or happiness in oneself or others, may not be as highly valued as time spent achieving success in the marketplace.

This is not to say that there is anything wrong with valuing success in this sphere. As cultural theorist Richard Sennett argues,[22] work is a valuable source of identity and self-respect. Those without

22   R. Sennett, *Respect in an Age of Inequality* (New York: W.W. Norton, 2003).

work are much less likely to be satisfied than those in employment. And, as sociologist Arlie Russell Hochschild argued in *The Time Bind*,[23] many individuals, faced with unfulfilling, unmanageable, conflict-ridden domestic lives, flee into the workplace where they can find the support, praise and gratification that they expected to gain at home.

But the value placed exclusively on paid time raises questions about how we regard time at home. For example, problems may arise if people believe in the concept of time as currency and feel that they should control time at home by creating rigid schedules which make it easier to 'manage'.[24] The reality of family relationships can rarely be tidied into neat little units spent productively achieving tasks. Nevertheless, for many parents the practice of valuing quantity over quality infringes on the way they 'run' their families. Parents put the kids in front of the latest Disney DVD so they can multi-task in another part of the house; and passive consumption replaces the time-consuming engagement of creating 'fun' things to do.

> Parents put the kids in front of the latest Disney DVD so they can multi-task in another part of the house

The key issue is not the way individuals value time: after all, many see childcare as one of the most important jobs they can do. It is how society values the way we spend our time. Time spent 'productively' in market-related activities continues to be valued over time spent outside the economic sphere. The primacy of paid work — often stemming from financial need, career aspirations, a desire for social status or just the fact that it is seen as interesting — means that many individuals feel they have to work long hours. And this then causes further difficulties as they do not have enough time at home, which can make time outside paid work more difficult.

23 A.R. Hochschild, *The Time Bind: When Work Becomes Home and Home Becomes Work* (New York: Henry Holt & Company, 1997).
24 Meikins, *op. cit.*.

3 Some time is better than other time

Time is an increasing preoccupation of governments, organisations and individuals because some time is seen as better than other time. It is not necessarily that we now have more hectic lives; after all, for centuries many people worked every waking hour doing paid work, laundry by hand or childcare among other things. But we now have far more choices about how to spend our time, making it more difficult to make decisions about how much time to spend doing a particular activity.

We are also arguably more judgemental about how our time is spent. One need only see stories in the *Daily Mail* criticising working mothers, and contrast them with the long working hours implicitly demanded by many organisations staffed by women who read the *Daily Mail*, to see that we are living out two contradictory ideals at once: spending quality time outside work as well as investing quality time in paid work.

And it is the idea of quality that is key. Money has a relatively stable purchasing power: a certain amount of money gives the power to purchase a certain quantity of any given product. Quality is irrelevant: a ten-pound note has no more quality than two fivers. But with regards to time, quality matters as well as quantity. Snatching 30 minutes of time with your children when shattered and grumpy does not amount to quality time. Working yourself into the ground in order to meet deadlines is not quality time. This is no way to build a sustainable, productive organisation, nor to ensure that individuals lead satisfying lives. Clearly each individual will have his or her own ideal mix of 'time'; and each organisation has different needs. But when we forget that one hour is not the same as the next, that the 13th hour of a working day is unlikely to be as productive as the second, then we fall into the trap of trying to financially engineer time, rather than use it effectively and enjoyably.

When we forget that one hour is not the same as the next then we fall into the trap of trying to financially engineer time

Time for change

Clearly, then, we need to change the relationship between time and money; to break down our assumption that time can be controlled, measured, and used as a economic variable. The idea that time is money, or that time in automatically equals productivity out, is far too simplistic. It was simplistic in the age of Henry Ford, when he realised that five days were better than six days; it is even more out of touch in a world that demands customisation and relies on knowledge, relationships and customer service.

We need to shift the debate from the 'time is money' principle to talking about time and quality. Organisations need productive time, not just lots of time. They need to ensure that focusing on the now doesn't jeopardise tomorrow. And if we focus on the quality of time, not just its length, we can start to engage with the changes in how we live our lives. We can start to respond to individual and organisational needs, to move from financial engineering to business building. Satisfied employees make a difference to productivity: in one retailer a 1% increase in commitment led to increased profits of £200,000 per month. It's about time we caught up with the 21st century and changed our approach. Not time is money, but time for change.

1

Taking people's time seriously

David Boyle

Let us give generously, in the two currencies of time and money.

Tony Blair, March 2000[1]

The originator of the phrase 'time is money' was the pioneer printer, philosopher, scientist and statesman Benjamin Franklin, who coined it in *Poor Richard's Almanac*, his most successful book of aphorisms published first in 1732. The phrase has alternately cajoled and irritated people for the quarter of a millennium since he wrote it.

Franklin had by that stage discovered the pleasures of turning his time into money in a very personal sense by printing it at home, and

---

1 Speech given by the Prime Minister, Tony Blair MP to the Active Community Convention and Awards, 2 March 2000.

watching the beneficial effects of a little extra paper money on the economy of Philadelphia. His newspaper, the *Pennsylvania Gazette*, used to apologise for not appearing on time because he was 'with the Press, labouring for the publick Good, to make Money more plentiful'.

The 'time is money' catchphrase was designed to be more puritanical: a maxim intended to prevent us from wasting one of those inelastic assets we are born with. It remains the kind of thought put into words by parents to their children, teachers to their pupils and managers to their staff, and as such has been relegated to the attic of political correctness.

It still grates rather. 'Time is money' has become self-satisfied and far too smug to fit neatly into the mouths of politicians. And yet, in another sense, it contains an element of truth that previous generations might not have fully unpacked. Time is indeed a kind of money. It is a birthright for us as individuals, and it is an extra dimension to any situation that can enable us – using other assets – to escape from poverty or misery or both.

All human beings have it as a right from birth. It is a kind of regular basic income of 24 hours a day, eight of which we have to spend asleep – a kind of tax which enables us to start relatively fresh and healthy the next day. Nobody has any more than 24 hours each day, and every morning we wake up with 24 more. It is a form of wealth that will never be cornered by the market and reminds us – and we need reminding – that not everything can be measured in terms of money.

This is not the place to discuss complementary currencies, but throughout history – and especially in the US since Franklin's days – there has been a healthy tradition of communities using different assets as media of exchange. It might be tobacco, it might be printed paper notes – these days denominated in local currencies, rather than in dollars, so that they are legal. It might even be frequent-flyer air miles, which were used by at least one airline until recently to pay their international consultants.

We can expect a range of new complementary currencies emerging which give value to the assets that people possess locally

All of these are used partly because conventional currencies are scarce and partly because they tend not to measure everything very accurately. And as the measuring qualities of conventional money becomes increasingly captured by the narrow interests of Wall Street and the City of London, we can expect a range of new complementary currencies emerging which give value to the assets that people possess locally – people's energy, wasted space, last year's computers, green fields. And of course people's time.

That is the idea behind using time as a kind social measuring system that can also reward people for the effort they put into their local neighbourhood.

In practice, this is known as time dollars in the US and as time banks in Europe. It is not strictly using time as a kind of money, because – like air miles and other reward points – time 'credits' do not have to balance at the end of the day: they are issued to measure and reward and, when they are redeemed, they just get deleted. On the other hand, it does mean giving renewed value to people's time – everybody's time, not just the financiers and brain surgeons. It also means treating people's time as a valuable asset that society needs.

Time banking

The story of time banks goes back to 1980, when the American civil rights lawyer Edgar Cahn first began planning how they might work. He was recovering from a major heart attack and the experience of feeling useless in hospital was profoundly influential, giving him a flash of insight into how so many other people in society must feel in different circumstances.

70% of participants suffering from a combination of physical and mental problems reported some remission of their condition within six months of joining the time bank

To develop the idea, he spent a sabbatical year at the London School of Economics in 1986, and was haunted by the English word 'redundant', which he heard for the first time on the BBC. The idea of his time 'credit' scheme was to make sure that people's time was valued — even if it had no market value. The result was a pamphlet called *Service Credits: A New Currency for the Welfare State.*[2]

The scheme was not taken up in the UK, and time dollars — or 'service credits' to give them their generic name — were pioneered instead by the Robert Wood Johnson Foundation, the biggest healthcare foundation in the US, in six pilot projects starting the following year. One of those, Member to Member, was based in the health insurance company Elderplan in Brooklyn and survives to this day, having become an important inspiration about what is possible on both sides of the Atlantic.

Member to Member now operates in different centres across all the boroughs in New York City, and was the inspiration for a similar programme in the UK, where a time bank was launched in the surgery of the Rushey Green Group Practice in Catford, south London.

Thanks to the time bank there since 2000, patients can now be prescribed not only with the usual pharmacological armoury but with a friendly visit once a week, or a lift to the shops. They can also be referred to the time bank — for example, in the case of long-term depression — if doctors feel that engagement of some kind would be useful.

The evidence is that this works. Elderplan has found that its customers stay clients for longer if they are participants in Member to Member, and there is evidence that they require fewer health interventions. Early research at Rushey Green showed that 70% of participants suffering from a combination of physical and mental problems reported some remission of their condition within six months of joining the time bank. There is confirmation of this in research by the Socio-Medical Research Group at King's College, London, which shows that those participants who are most actively involved in the time bank experience the most improvement in both their mental and physical health.

2   E.S. Cahn, *Service Credits: A New Currency for the Welfare State* (Discussion Paper 8; London: Suntory-Toyota International Centre for Economics and Related Disciplines; London School of Economics, 1986).

In both Brooklyn and Rushey Green, organisers used the idea of a DIY scheme — training up relatively elderly people to do small repairs, sometimes as little as changing a light bulb or putting up curtains — to provide an urgent service to older householders, and at the same time engage more men in the time bank. In Brooklyn, one of the oldest of these, wearing a 'DIY team' cap, featured in the most recent Elderplan advertising campaign with the slogan: 'Does your insurance company give you a friend like George?' Experience with both is that the time bank is able to provide the kind of human, face-to-face services that professionals have no time for and perform less well than neighbours.

The time bank is able to provide the kind of human, face-to-face services that professionals have no time for and perform less well than neighbours

The Blairite sociologist Professor Anthony Giddens was among the first in the UK to propose time banks based on the original American time dollars idea. 'Volunteers who take part in charitable work are "paid" in time donated by other volunteer workers,' he explained in *The Third Way*.[3] 'A computer system registers every "time dollar" earned and spent and provides participants with regular accounts. Time dollars are tax free and can be accumulated to pay for healthcare as well as other health services.' Cahn himself described the idea as more like a blood bank or babysitting club:

> Help a neighbour and then, when you need it, a neighbour — most likely a different one — will help you. The system is based on equality: one hour of help means one time dollar, whether the task is grocery shopping or making out a tax return ... Credits are kept in individual accounts in a 'bank' on a personal computer. Credits and debits are tallied regularly. Some banks provide monthly balance statements, recording the flow of good deeds.[4]

But, over the last decade, the idea became more sophisticated than that. Participants' efforts are rewarded with 'time credits',

3 A. Giddens, *The Third Way: The Renewal of Social Democracy* (Cambridge, UK: Polity Press, 1998).
4 Cahn, *op. cit.*.

which they can spend on help when they need it themselves, but also on a range of other things that recognise their effort even if they don't — by necessity — fully 'pay' for people's time. These can include tickets to local sports matches (Gloucester) or entry to local sports centres (Peckham), but more usually they can sometimes make participants eligible for a refurbished computer to take home or for training.

The idea of giving away refurbished computers for time credits emerged in Chicago in a highly successful scheme to engage disaffected pupils in inner-city schools as tutors for younger pupils. The credits they earned were able to go towards the computer. Their parents were also expected to earn four credits, either by helping out in their local school or by attending police–community liaison meetings.

It was continued in Washington, DC, in one of the most innovative projects of its kind, which now enables between a quarter and a half of all first-time young defendants in the city to be taken out of the exhausted courts system and to be tried by juries of other teenagers, who 'sentence' them to appropriate training or rehabilitation or some other activity to atone for their actions. The jurors, and offenders on community service, earn time credits that they can use to buy computers.

By then, time banks had made the shift to the UK after a very successful visit made by Cahn in 1997 to London and Newcastle. His message — that the so-called 'problem people' have assets that helping professionals need — was warmly received on that visit and on subsequent visits, all of which were made possible with the support of healthcare foundation the King's Fund. Stonehouse was the first of a network of nine time-based currency projects planned for Gloucestershire, thanks to a series of grants to Fair Shares the following year, from the Barnwood House Trust and the Community Fund. The next project opened in Newent, Gloucestershire, in 1999, followed by four linked time banks in Gloucester itself.

Rushey Green, HourBank in Peckham — with its highly successful regular community café bringing together asylum-seekers and locals — and the Angell Town Time Bank in Brixton, then formed

Time banks facts

- Time-based currency schemes use time rather than money in exchange systems which reward people for the effort they put into their community. They are known as **time dollars** in the US and as **time banks** in Europe.

- According to the Time Banks UK website, time banks 'make it easy for people to get involved and to make a difference. They encourage community participation and actually reward participants, and so strengthen community activities.'[5]

- American civil rights lawyer Edgar Cahn first came up with the concept of time banks in 1980. In 1986, after a year-long sabbatical spent developing the idea, he published *Service Credits: A New Currency for the Welfare State*.

- These 'service credits', better known as time dollars, were pioneered by the Robert Wood Johnson Foundation, the biggest healthcare foundation in the US, in six pilot projects starting in 1987.

- In the UK, Blairite sociologist Professor Anthony Giddens was among the first to propose time banks based on the original American time dollars idea.

- In 2001 London-based think-tank the New Economics Foundation launched the London Time Bank network, which is now nearly 40-strong.

- Computer programmer Kent Gordon distributes the free *Timekeeper* software he created to help people set up and manage their own time banks.

- There are around 120 time bank programmes around the UK. China and Japan also have time banks, and similar networks in Spain and Italy. The schemes are typically adapted to local issues and vary from country to country.

- The HourBank in Peckham, London, regularly runs a highly successful community café bringing together asylum seekers and local residents.

- In one community currency scheme in Japan, participants receive 20 stones to spend at the beginning of each year, with one stone being equivalent to one hour.

- Time banks sometimes offer goods and services to participants who might not otherwise be able to afford them. In Chicago, disaffected pupils in inner-city schools receive refurbished computers in exchange for time credits.

- In Washington, DC, one of the most innovative projects of its kind enables a quarter to a half of all first-time young defendants in the city to be taken out of the courts system and tried by a jury of other teenagers, who 'sentence' them to appropriate training or rehabilitation or some other activity to atone for their actions. The jurors, and offenders on community service, earn time credits that they can use to buy computers.

- Doctors at Rushey Green Group Practice in Catford, south London, prescribe some patients with visits from time bank volunteers, as well as participation in the time bank scheme. Early research showed that 70% of participants suffering from a combination of physical and mental problems reported some remission of their condition within six months of joining the time bank.

- Research by the Socio-Medical Research Group at King's College, London, showed that participants who were most actively involved in the time bank experienced the greatest improvement in their mental and physical health.

- In 2003, time banking in the UK hit the 100,000-hour milestone. In just five years UK time banks exchanged 100,000 hours of time credits—the equivalent of 50 years of full-time work.

- 'The Time Bank model provides a simple solution with groundbreaking results. We are on a journey to find how individuals can create a better world that relies not just on financial support, but also on each other.' — Time Banks USA

Further information is available at:

www.timebanks.co.uk — Time Banks UK

www.timebank.org.uk — national time bank campaign

www.timekeeper.org — *Timekeeper* software

the kernel of the London Time Bank network, which is now nearly 40-strong. The network was launched by the New Economics Foundation in 2001, and has encouraged a range of public- and voluntary-sector organisations to employ time bank development officers to spread the idea as part of their community engagement strategy — including Lewisham, Islington, Southwark and Lambeth borough councils, the South London & Maudsley NHS Trust and the Hexagon Housing Association.

There are now around 120 time bank programmes around the UK, with rather more in China and Japan, and similar networks in Spain and Italy — though each country has adapted the idea to tackle slightly different issues.

Time banks as social glue

The way time banks have developed on both sides of the Atlantic has been to shift away from the free-standing infrastructure that enables reciprocal and informal volunteering, and towards a technique — based mainly in public services — that is able to measure and reward the effort people put in to make their neighbourhood work. Time banks, in other words, are not so much a rival form of volunteering, but a social glue that can draw together local projects and help them to reach the groups that never normally take part. It can be said to build social capital: the resources and connections we share as a society. Research by the University of East Anglia suggests that time banks are uniquely able to access support from some of the hardest-to-reach groups in society. Over the first quarter-century of the idea, the following lessons seem to be increasingly clear:

1. **Everyone has something to offer.** Time banks work because there is no artificial division between givers and receivers, and because there is an ethic at the heart of them

— that everyone will be asked to give something back. It was clear as early as the Brooklyn project in the 1980s that not only did the bedridden elderly welcome being asked to contribute — maybe by making supportive phone calls to neighbours but it improved their health to do so.

2. **Feeling useful is a basic human need and can be transformative.** Many participants are people who have spent their whole lives defined (by professionals) by their disabilities, and they are never asked for anything back. By defining them according to what they can do, time banks seem able to transform people's lives.

3. **Systems work through face-to-face contact.** Success or failure depends very much on the time broker at the heart of the scheme, and how much they can push or trust participants. Time banks are not primarily the software — useful as that is: they work because they bring people face to face with each other, across age, race and other cultural divides.

   Time banks are local schemes and have to remain so — people do not volunteer for their local authority, still less for the government

4. **They have to be local.** That basic informality, valuing everyone's time the same, makes all the difference. Time banks can perhaps be accredited centrally, but they are local schemes and have to remain so — people do not volunteer for their local authority, still less for the government. Time banks work because they are human-scale.

But the key lesson is that time banks can create reciprocal relationships between people and institutions, as well as between people and people, which ordinary volunteering finds it harder to achieve. They allow almost anybody in society, including the elderly and housebound, to give something back. And the evidence is that feeling needed is a critical missing piece of the social capital jigsaw.

That is what makes time banks the critical tool in transforming clients of public services from passive supplicants to active — even equal — participants alongside professionals in the business of delivering care or education, or tackling crime. The evidence is that this is the critical element in their success or failure. People's time, it turns out, is extremely valuable.

That is why time and 'co-production' — as the philosophy that underpins it is coming to be known — have found themselves at the heart of the debate on both sides of the Atlantic about welfare, philanthropy, public services and why they are often so intractable. It is why those involved in the debate have been asking how, after nearly six decades of the welfare state, we seem to have made so little difference to poverty, youth crime, ill-health and school failure.

The vital role that ordinary people can play in crime or disease prevention has been made clear by research on both sides of the Atlantic for decades now. Like the dramatic 1997 findings by the Harvard School of Public Health in its study of more than 300 neighbourhoods of Chicago — that the key determinant of the crime rate isn't income or employment, but trust: whether or not people are prepared to intervene if they see local children hanging around.

The underlying message behind this and other findings was the vital importance of extended relationships, trust and informal net-works. And even more important: that the time of patients, old people, neighbours, busybodies, are also vital assets — necessary to preventing crime, keeping people well, bringing up children and all the other tasks that society currently struggles with. The difficulty is that our welfare systems and philanthropic bodies are geared in the opposite direction — defining clients primarily by what they lack.

Co-production first emerged at the University of Indiana in the 1980s as a way of explaining how policing fails without community support. But the concept has since been refined and extended by Cahn, accusing professionals of creating dependency — and a dependency of a peculiarly corrosive kind: one that convinces clients they have nothing worthwhile to offer, and which under-mines what systems of local support do still exist.

Public services that effectively treat the time of their clients as a resource will be organised very differently. That is why American professionals who flirt with co-production are often treated with suspicion by their colleagues. The founder of the charity Home-builders, child psychologist Jill Kinney, had built an organisation devoted to sending professionals to work with at-risk families rather than take children into care. When Jill publicly questioned whether the permanent neighbourhood support that families needed might be provided by neighbours once the professionals had gone as they have to go eventually — and by other lay people who had faced similar problems themselves, she was banished by her own organisation. Her new organisation, 'Home, Safe', provides just the service she was advocating

But sometimes the issues are so intractable that there is really no alternative to getting ordinary people to do what had once seemed the sacred preserve of professionals. In 1994, one of the local workers employed by the US charity Partners in Health in Lima died from MDR-TB (multi-drug-resistant tuberculosis). It soon became clear that hundreds of locals were also suffering from MDR-TB, thanks to disastrous treatment programmes in the late 1980s.

Medical opinion now says that MDR-TB requires such expensive drugs — and such complicated safeguards to make sure that courses of the few powerful antibiotics are completed — that only the very wealthiest communities can afford to start. Health agencies advise developing countries not even to try.

But, against their advice, Partners in Health solved the monitoring problem by training the local community to supervise the drugs in patients' homes. Local people also designed individual treatments to suit each patient, with great success. Partners is now achieving cure rates of 80% — as good as anything achieved in the US but at a fraction of the cost, and by using people's time as a resource — and have brought the lessons home to Boston.

Cahn himself first used the 'co-production' phrase to explain his approach to training lawyers. Students at the District of Columbia School of Law are trained on the job by providing legal support for people and communities who need it but can't afford it — and this is where co-production comes in, because they don't do it for free. They charge out their time in 'time credits'. This is not philanthropy; or, if it is, then it is philanthropy of a whole new reciprocal kind. The recipients of legal advice pay off their bill either by passing on what they have learned to somebody else or by helping out in the community in some other way.

This kind of thinking is a partial explanation for why neither more professionals nor more money seems to make enough difference to our welfare systems: the reason is that neither fully values the time of lay people or clients. Without using this time, more professionalism and more money might even perpetuate the original problem.

This has some profound implications. It means, for example, that philanthropists would no longer simply give away their money or expertise — they would trade it. Not for money, of course, because the neighbourhoods they are helping don't have it, but for their time. It means the main focus of welfare is not the failures of the person before you at the desk, it is their capabilities and how they can put them to use.

> Neither more professionals nor more money seems to make enough difference to our welfare systems: the reason is that neither fully values the time of lay people or clients

It also raises another question about society: if people are using their time to make public services work, or to deepen them, or to make their neighbourhoods liveable — should they not be given the basic necessities of life? In other words, should not whoever is responsible back people's time credits — even though they may never hold down a conventional marketable job — with the food, shelter, clothing and companionship they need?

Towards a new mutualism

That is a glimpse of a new kind of mutualism — one that takes everybody's time seriously and accepts that society needs people to use some of their time on its behalf. We know that traditional forms of mutualism failed to motivate people, and perhaps that's not surprising considering how little they felt involved. We know the new forms of 'consultation' are not working, by themselves, and are certainly not giving people a sense of ownership.

This new form of mutualism — co-production that takes people's time seriously — has participation at its heart, and a broader definition of productive work. Mutual participation without ownership can be exploitative, of course, but mutual ownership without participation is meaningless. It's not the owning, in other words, it's the taking part. Because only by investing our time as equal partners in our institutions do we get any kind of meaningful control over them.

In that way — taking everyone's time seriously — we might expect a little more of our welfare services, more than the current rather hopeless sense of maintaining a miserable status quo. Because by taking everyone's time seriously, they might actually work.

8

Ethics in time

Mary Warnock

One of the great distinguishing marks of human as opposed to other animals is that our brains are so organised as to be capable of what we call imagination. I define imagination, following Jean-Paul Sartre, as the ability to contemplate what is *not* as well as what is. This means that human beings are not confined to the here and now, what is before our eyes, ears and noses. We may think about how things might have been as well as how they are: for example, 'suppose I had painted my room blue rather than green?' We think how things used to be (the past) and how things probably will be (the future).

Other animals remember, but as far as we know their memory is usually, at least, triggered by a present event. A horse that has once been frightened by a piece of paper blowing in the wind at a particular spot may see spooks there every time it goes past; or it may recognise a horse-box as another of those things it is reluctant to enter. It *may*, of course, be able to spend its declining years running over again its triumphant races of the past; but alas, because it has no language, we cannot know. Our ability to talk is intimately bound up with our ability not just to recall but to imagine the past. And, similarly, though with less security, we can envisage the future. You could say that other animals remember and foresee but with no sense of time. We, on the other hand, in all our serious thinking are inescapably caught up in our sense of time.

Philosophers and the future

Nowhere is this better exemplified than in our thinking about ethics, about right and wrong, good and evil, both theoretically and at a practical level. Most obviously this is evident in the thinking of utilitarians, philosophers who hold that the distinction between right and wrong must be drawn, if it is to make any sense, from a consideration of the consequences of our actions, and whether these consequences will be harmful or beneficial to people in general. To make such distinctions we must be capable of seeing into the future. Utilitarians, whether philosophers or politicians, must consider an action and ask themselves what it will lead to; or they must consider a *kind* of action, such as deceiving children about their paternity, and ask what is the likely consequence of such actions on children in general. Indeed, one common objection to utilitarianism as an ethical theory is that it demands the impossible: no one can trace the consequences of an action, or even a kind or type of action, into an unending future.

> We in all our serious thinking are inescapably caught up in our sense of time

However, it is not only utilitarians whose ideas of good and evil are inextricably linked with time. That great philosopher of the 18th century Scottish Enlightenment, David Hume, distinguished between the Natural and the Artificial Virtues.[1] Natural virtues were those towards which men, and indeed other animals, had an

---

1   D. Hume, *A Treatise of Human Nature: Being an Attempt to Introduce the Experimental Method of Reasoning into Moral Subjects* (1739-40).

instinctive propensity, such as love of their children and a desire to endure hardship in order to protect them. Artificial virtues, by contrast, were those which human beings as social animals would always — now and in the future — perceive as essential to prevent society's descent into chaos. They included virtues that other animals could not conceivably practise, such as promise-keeping and a regard for justice. These were the human virtues from which the concept of a *moral principle* — or duty that could override one's immediate inclination — could be derived.

Hume's contemporary in Prussia, Immanuel Kant, also identified a moral imperative as that which, like the laws of logic, was time-less.[2] Kant thought that any consideration of the consequences of a morally good action was out of place and that duty, including justice itself, was something that had to be maintained, as he famously expressed it, 'though the heavens fall'. The obligation to keep a promise, for example, was not dependent on the circumstances in which one happened to find oneself. It was absolute, and a consideration of now and the future was irrelevant. Morality, unlike all other aspects of human thinking (with the exception of logic and mathematics), was specifically placed *above* or *outside* time; and timelessness is a notion that manifestly makes no sense except to one who has accepted the idea of time in other aspects of human life.

> Morality, unlike all other aspects of human thinking (with the exception of logic and mathematics), was specifically placed *above* or *outside* time

Such an idea of the timelessness of morality is, of course, even more familiar in the context of religion than that of philosophy. It is perhaps especially central to Judaism, a religion totally grounded in a sense of history, and the contrast between the everlasting and the ephemeral:

> But I said, O my God, take me not away in the midst of my age: as for thy years, they endure throughout all genera-tions. Thou, Lord, in the beginning hast laid down the foundation of the earth: and the heavens are the work of

2   I. Kant, *Groundwork of the Metaphysics of Morals* (1785).

they hands. They shall perish, but thou shalt endure; they all shall wax old as doth a garment; And as a vesture shalt thou change them: but thou art the same and thy years shall not fail (Psalm 102, *Book of Common Prayer*).

And it is therefore, the Psalmist expects, from this God that the laws of morality are to be learned.

So it seems to me that the idea of time is woven into the idea of morality, the attempt to do good rather than evil, not just accidentally but of necessity. But of course in the real world, and especially in the world of political decision-making, we are all to some extent at least utilitarians. A decision of public policy, unless taken by a tyrant indifferent to the people it will affect, is bound to consider the effects of the policy on people in general. Will its outcome be productive of more good than harm?

> In the world of political decision-making, we are all to some extent at least utilitarians

I shall illustrate the difficulties with which policy-makers are confronted by two recent examples. The first is drawn from the field of bioethics, the second from criminal justice.

Science and change

The development of science and technology in the field of bioethics is extraordinarily rapid. The more exciting the developments, the greater the number of bright scientists who want to be involved in the research, and thus innovation accelerates. As a matter of logic, it is impossible to foresee creative, original and imaginative developments, whether in the arts or the sciences (for, if one could, they would hardly be original). However, whereas in the arts it is possible simply to wait and see what originality brings forth, with the sciences it is different. For physics, chemistry, biology and medicine have practical outcomes, certain to be morally significant. It is therefore necessary to frame a public policy that will take into account and perhaps regulate what will happen next, as well as what is happening here and now.

At present, in medical practice, the abortion of foetuses found to be seriously defective is a reality; it is permitted by law, as is selective implantation of embryos fertilised *in vitro* (i.e. in the test tube), where there is a high risk that a child born to the parents will suffer a genetically inherited disease. In some cases the embryos are screened for the condition, and only those unaffected chosen for implantation; in others, where only male foetuses are at risk (as is the case with haemophilia) only female embryos will be implanted. However, in the future there lies the possibility of genetic manipulation of the embryo *in vitro* or even *in utero* (in the womb). The genes of the embryo could be altered so that, when the baby was born, it would be of the preferred sex; or it would be tall and blond; or it would be possessed of extraordinary mathematical powers, or musical genius, or amiable temper. The more we learn about the human genome, and the way genes interact with one another and are expressed in certain abilities or patterns of behaviour, the more it seems possible that such interventions could actually occur, outside the realms of science fiction.

Many people are alarmed that the skill and knowledge that would make such genetic engineering possible would have terrible consequences. So should we nip such future developments in the bud? Should we legislate to prevent any genetic alteration to an embryo, or a child who has been born? Should even sex selection be prohibited except to prevent the birth of children at risk of inheriting certain specified serious diseases?

It will be obvious that this example is one among many possible examples in the field: should the reproductive cloning of human beings be prohibited before it has even been attempted, on the grounds that a human clone would be a kind of monster? Should the genetic modification of crops be prohibited because of its presumed effect on the environment and biodiversity? In all these cases Parliament, and society, on whose behalf it legislates, is being asked to decide whether the consequences of allowing the new techniques to develop unregulated will be so disastrous that regulation or even prohibition is a moral imperative.

Future scenarios

> Those who have knowledge don't predict. Those who predict
> don't have knowledge.
>
> *Lao-Tzu*

From stock market tips to Nostradamus, since the dawn of time people have tried to predict the future — with varying degrees of accuracy. We pay astrologers and psychics a fortune every year. Some organisations such as BT employ futurologists, people whose job it is to imagine life decades, even centuries, from now.

In fact our very nature, our extraordinary intelligence, evolved to help us imagine future outcomes: what will happen if I eat this poisonous plant, for instance, or chase that promising animal? Our capacity to envision and to predict is the tool that has allowed humanity to develop from a savannah primate to a tool-using, gun-toting, environment-changing species which inhabits almost the entire globe. Unsurprisingly, then, future scenarios pervade fiction and business alike, with important repercussions for sustainable development.

Scenarios, describing different possible futures and the pathways towards them, can be highly useful for large-scale decision-making. Governments and multinational corporations use them to study long-term trends in areas of high complexity and risk. During the Cold War the defence industry began to develop scenarios for use in their military strategy. Then in the 1970s Shell applied the same model to business planning because of the difficulties it was facing in the oil crisis. Peter Schwartz, who worked for the oil giant at the time, went on to write *The Art of the Long View*,[3] the seminal text on scenario planning.

Shell, however, famously overestimated the oil reserves it had left, causing rather unfortunate problems down the line. As Shell chairman Philip Watts explains in *Exploring the Future*,[4] 'Scenarios are not prophecies or preferences. They are . . . credible alternative stories about the future . . . designed to help us challenge our assumptions

---

3 P. Schwartz, *The Art of the Long View* (New York: Doubleday, 1991).
4 P. Watts, *Exploring the Future* (Global Business Environment; Shell International, 2003).

. . . and test our strategies and plans.' In other words, because we are active players in defining what will happen, and because we have an imperfect grasp of the factors that determine the future, scenarios are most productive in highlighting growing problems, and in envisioning a future to aim for — or to escape from.

The WBCSD (World Business Council for Sustainable Development) imagines three different scenarios for the future up to the year 2050:[5]

- FROG!
- GEOpolity
- Jazz

FROG! stands for 'first raise our growth!', the imagined cry of developing nations protesting that sustainability should take the back seat to economic growth. FROG! warns that unless we take action now, on a global scale, we may not heed the signs until it is too late.

In GEOpolity 'people begin to look for new leaders and to demand new social institutions' — such as sustainable cities, and industrial ecology, 'even if doing so requires economic sacrifice'. Governments, under pressure from the global citizenry, 'take the lead in shifting the structure of the economy towards sustainable development'.

Dynamic reciprocity drives the world of Jazz, where 'diverse players join in ad hoc alliances to solve social and environmental problems in the most pragmatic possible way'; where 'NGOs, governments, concerned consumers, and businesses act as partners — or fail'. In this scenario, environmental and social values are incorporated into market mechanisms rather than being enforced by official institutions, as in GEOpolity, or ignored, as in FROG!

The WBCSD encourages us to 'practice a flexible approach to the future' — to consider it from different points of view, remembering that it is shaped by a complex interplay of forces and events.

One need only look back on past predictions to see the weakness of making guesses about the future, no matter how educated. In 1903, the *New York Times* proclaimed confidently that human flight was impossible — just one week before the Wright brothers flew at Kitty Hawk, North Carolina. In the 1950s, electricity companies optimistically declared that energy would soon be so cheap it couldn't be

5 WBCSD, *Global Scenarios 2000–2050: Exploring Sustainable Development* (Geneva: WBCSD).

measured; and in 1981, Bill Gates said that '640k of memory ought to be enough for anyone'.

That isn't to say that all visionaries get it wrong. Leonardo da Vinci is famous for describing and sketching inventions that weren't built until hundreds of years after his death, such as the helicopter, parachute and tank, among others. And science fiction writer Arthur C. Clarke correctly predicted the development of the geosynchronous communication satellite.

Other authors have imagined both utopias and dystopias, penning metaphors on a grand scale, visions of a world that grapples with sustainable development — and sometimes fails. *Stark*, by Ben Elton,[6] sees the powerful and wealthy attempting to exit the atmosphere in a private spaceship, well aware that they have plundered the earth to the point of provoking an environmental catastrophe. George Turner sets his novel *The Sea and Summer*[7] in Australia at a time when humanity is suffering ecological, economic and social collapse (in the years 2041–61).

Ursula K. LeGuin uses fiction to test social theories in *The Dispossessed*,[8] which describes a rebel society built on the principles of equality and justice where citizens freely share work and possessions alike. Numerous films such as *Waterworld*, *The Day after Tomorrow* and *A.I.* envision the earth after a climate change holocaust; but arguably they have not done much to bring sustainability onto the 'public's' radar, perhaps even encouraging fatalism. Despite their weaknesses, scenarios in business and fiction have their good sides, encouraging us to take the future, and therefore sustainable development, seriously. As for proposing solutions — scenarios can only tell us what might happen, not what to do about it. That, and the future, is up to us.

Chris Sherwin and Mireille Kaiser

6 B. Elton, *Stark* (Time Warner Paperbacks, 1989).
7 G. Turner, *The Sea and Summer* (Grafton Books, 1989).
8 U.K. LeGuin, *The Dispossessed: An Ambiguous Utopia* (Avon, 1974).

Those of a sanguine disposition (among whom I, with reservations, include myself) are inclined to say, 'Let's stick with what happens now. Let's see whether what is now permitted is harmful, before rushing ahead to fend off imagined future horrors.' But many argue that, since these are moral or ethical issues, raising questions about what kind of society we want to live in, it is absolutely necessary to decide what we *would* say if the future caught up, if fantasy became reality: if, for example, it were possible to clone human beings. It is not enough to say that the development you envisage is not yet technically possible. You must be forearmed with a principle to apply when need arises.

Gyges's ring and the slippery slope

I do not deny that we are quite accustomed to hypothetical moral thinking. Plato introduced such a device in his philosophical treatise on governance and the moral life, *The Republic*. The Ring of Gyges was a mythical object which, when worn, rendered its wearer invisible. Plato asked whether, if we had such a ring, we would go round committing crimes in the sure knowledge that we would not be detected, or whether, as he believed, there is a kind of internal restraint (which we might call conscience) which would prevent our doing wrong, independent of the fear of discovery.

> It is not enough to say that the development you envisage is not yet technically possible. You must be forearmed with a principle to apply when need arises

The lesson of this thought experiment is in asking what we would do with something we don't yet have that enables us to do things that at present we cannot, however fantastic: it forces us to ask how we will react in the future. One reaction to this question is to invoke the idea of the Slippery Slope. Indeed it is the main force of the argument of those who advocate anticipating the worst by applying the precautionary principle.

The Slippery Slope argument goes thus: there may be little or nothing wrong with what happens at present (for example, selecting embryos of the desired sex, when the 'wrong' sex would carry the risk of serious disease); but unless a barrier is erected, the next step will inevitably be gender selection for frivolous reasons, and then genetic manipulation to produce desired characteristics, and then . . . and then . . . and so on. This often includes the fear that what society considers morally acceptable will shift, that a moral absolute will become morally relative. Who can say what horrors of eugenics and 'designer babies' lie at the bottom of the slope?

The barrier across the slope may take the form of prohibitive legislation (as has been introduced to prevent human reproductive cloning, in the UK and most European countries) or of a system of regulation, licensing and inspection. In either case, the purpose is the same – to put an end to the inevitable descent.

The metaphor of the slope is not just a spatial but a temporal metaphor. The slope is steep but also *slippery*; and this description has a dimension that is time-related. It entails a slide to the bottom that takes time, but all too little time. You are away and at the bottom before you either wish or know. The effective word in the metaphor is 'inevitable'. There is no *logical* necessity to connect taking the first step (say, therapeutic cloning) to the next and next (until we reach reproductive cloning on demand). It is more a reflection on human nature, a supposedly timeless entity, like God, that will never change. If human beings have one thing – so the argument runs in the abstract – they will want more; and more and more. Give them an inch and they will take a mile. They are greedy for power, knowledge and profits; and they need to be halted in their tracks. This, it is argued, is the only ethically respectable thing to do. The concept of time, and accelerated time seems, in the field of rapidly advancing science and technology, to demand the precautionary principle to save us from human nature.

> Who can say what horrors of eugenics and 'designer babies' lie at the bottom of the slope?

The end of trial by jury?

My second example is rather different. Its emphasis is more on
consciousness of the past than of the future. In the Criminal Justice
Bill 2003, it was proposed by the UK government that for certain
cases trial by jury should be abolished. These cases included, among
others, those trials for relatively minor offences where previously
the accused could choose whether their case should be heard before
a jury or not; and those lengthy and complicated trials for fraud
where, it was held, the jury was required to sit for months on end,
and where in any case the issues were too complex for them to
understand. This part of the Bill was most vigorously attacked and
finally thrown out by the House of Lords, including those judges
and retired judges who sit, but not by them alone.

The government's argument was based on the fact that there was
(and is) a huge backlog of cases waiting to be heard; and that trial
by jury in the cases in question constituted an enormous waste of
time, in the trivial cases often requested by the accused simply
because time could thereby be wasted. In fraud cases, the argument
was that not only was trial by jury a waste of time (the jury's as well
as the court's) but a waste of money as well if jury members had to
be compensated for lost earnings over many months. And, to crown
all, it was futile, since the jury could contribute nothing to the
verdict.

The argument of the majority in the House of Lords, on the other
hand, was that our system of justice had been built up over the
centuries on the availability to every accused person of trial by jury,
and that to remove this right from the accused simply in order to
speed up the processes of the courts was blatantly to prefer expediency to justice. Moreover, to suggest that members of the jury
could not be brought to understand complex cases was to undermine the whole basis of trial by jury, according to which those
standing trial should be heard, and their verdict finally determined,
by ordinary people, randomly selected. It was keenly felt, and
passionately argued, on all sides of the House, that a system of
justice that had been built up over years, and was commonly

regarded as the best way to ensure a fair trial, could not properly be abolished, or partly abolished by a government ruled by the short-sighted desire to save time and money. Nobody denied that there were inconveniences, sometimes severe, involved in the jury system, nor that it was often inefficient. But to allow trial by jury was a matter of principle and must be upheld as such.

There may have been an element of the Slippery Slope argument here (if *this* principle were overthrown, which would be next? Where would it end?). But, on the whole, the opponents of the government were not concerned with other, worse, things that might follow from the government's proposals. They were concerned simply that the general principle should not be allowed to lapse in those cases here and now where it was proposed that it should be abandoned. As we saw at the beginning, principles are, in a sense, timeless. Expediency on the other hand is concerned with today and tomorrow. And judges, of all people, are professionally devoted to the distinction between principled justice and what is immediately attractive or convenient, or what might get those who administer the courts out of a perhaps temporary administrative hole, or what would save money within the system. Many other examples could be adduced where ethical arguments seem indistinguishable from arguments about time, perhaps particularly where what is at issue is the value that we attach to the conservation of the environment. Good versus evil is almost equivalent to long-termism versus short-termism.

> Good versus evil is almost equivalent to long-termism versus short-termism

Experience and aspiration

It is easy to see how a belief in the importance of the past in making moral decisions could lead to a stultifying conservatism: 'Let there be nothing new.' Equally, the claim that the moral principles on which we base such decisions are timeless may seem unduly preten-tious, in the light of what we know about how such apparently timeless and changeless principles not only vary over geographical space but also change over time. It is part of the human predicament that in making moral decisions (especially decisions of public policy) we are poised uneasily between knowledge of the past and aspirations and hope for the future. It falls to human beings, as it does not to other animals, to make moral decisions; and only they are so poised between history and hope.

Perhaps we ought to remember two things: first, our aspirations for the future will get nowhere, indeed our society will collapse into chaos, if we too readily consent to abandon the long-term for the short-term, principle for expediency. Secondly, the particular situation we are now in, where we have to make decisions of policy about biotechnology, and how much intervention to allow in the processes of life itself, is new. In this situation we must not be too timid. We must not forget the good that we hope for from the new interventions, the gradual diminution of the power of disease, the alleviation of both hunger and poverty throughout the world.

> We must try to avoid the kind of conservatism that would deprive the world of the great benefits that we can half foresee from the genetic revolution. But, in doing so, we may have to re-examine and perhaps adjust some of our concepts of time

In a largely secular society, it is understandable that people cling to the idea of Nature as they once clung to the idea of God. And in our post-Darwinian way, we may tend to think of evolutionary time as nature's time, and to fear and distrust the new genetic science, which seems to hurry us along the path of change faster than nature intended. We must try to avoid the kind of conservatism that would deprive the world of the great benefits that we can half foresee from the genetic revolution. But, in doing so, we may have to re-examine and perhaps adjust some of our concepts of time. We need both to be a bit sceptical about claims of timelessness, and yet to weigh up seriously the values of the past against those of the imagined and not wholly foreseeable future. This has always been the great difficulty in framing public policy. Now, because of the spectacular advances of science, it seems more taxing than ever before.

9

Time and technology

James Goodman and Britt Jorgensen

In 1965, Gordon Moore, co-founder of the giant computer chip manufacturer Intel, observed that the number of transistors on computer chips, and hence computer processing power, had doubled every 18 months for the previous five years. He predicted that this would continue into the future, and it has. The number of transistors on a chip has risen in the 40 years since his eureka moment from 2,300 to 55 million.

This has happened through a series of small technological breakthroughs. One was to use copper instead of aluminium as the conducting material, increasing chip efficiency by 30%. Another was to use light to carve the circuits joining up transistors. Shortening the wavelength of light used means finer and finer circuits can be etched, which means that more transistors can be squeezed onto a chip.

The fastest computer in the world today is the Earth Simulator Centre, built in Yokohama, Japan, by NEC in 2002 and used to calculate possible futures for the planet's climate. It has a peak speed of almost 36 teraflops, which means that it can carry out 36 million

million million mathematical operations in a second, something that would take a person with a calculator 2.1 million years to achieve.

But what has come to be known as Moore's Law is up against serious physical limitations. One problem is heat. Chips generate an awful lot of it, which is why desktop computers have fans and why server farms — banks of server computers where much of the Internet's data is housed — use so much electricity that they sometimes cause power cuts in Silicon Valley. Another problem is that transistors are so tightly packed together on chips that it is increasingly difficult to control the flow of electricity between them, without electrons jumping about unpredictably.

So within current technological bounds Moore's Law is thought likely to hold true for another decade. But who's to say that within the next ten years the critical breakthrough is not made into another realm of technology, and quantum or laser computing becomes possible?

The world of computer chip manufacturing, server farms and microscale electronics may seem far removed from everyday life, but for the majority of people in the developed world, whose lives are entwined with technology, faster computers are another factor in the trend towards faster lives.

Concerns about the detrimental effects of technology speeding up society go back a long way. In 1907 the French author Paul Adam worried about the impact of the bicycle. The pedal bike allowed people to move themselves around four times faster than walking. Adam feared the emergence of a 'cult of speed' for the bicycle generation, and doctors warned that too much cycling, especially into the wind, would lead to permanent facial disfigurement.

Today, many understandably fear that faster technology, while creating step-changes in business efficiency and stimulating economic growth, is also accelerating the planet's environmental decline, speeding up climate change and loss of biodiversity — which is ironic when the world's fastest computer was built specifically to understand climate change. The faster we live, the faster we consume the planet's finite resources and trash the natural systems on which we depend. In this scenario, increasing speed is only driving us faster towards self-destruction.

At a personal level the effect of faster lifestyles is more stress. A survey carried out at the end of 2002 by Essex University suggested that more than a third of employees in the UK felt overworked.[1] In 2003, stress replaced back pain as the number one reason for absenteeism. A quarter of British families share a meal together only once a month. With always-on communications keeping us permanently in touch, Western society is moving more and more towards a 24/7 approach to life, enabling us to work more, do more and consume more. Even the term '24/7' is a new, quicker way of saying 'all day every day' or '24 hours a day, 7 days a week'.

A quarter of British families share a meal together only once a month

Losing control of time

Much of the stress that we feel in our day-to-day lives stems from a feeling that our time is not really *our* time. We have no control over it and, despite our best efforts, have become slaves to the unrelenting onward march.

There's a paradox here. A lot of the technological tools we use at work, at home or on the move are designed to help us manage our time. The PDAs (personal digital assistants) that we carry around with us have calendars and diaries that can co-ordinate wirelessly with the calendars on our mobile phones and desktop computers. They will sound an alarm if we are late for an appointment, or warn us 15 minutes in advance, to give us time to prepare. Network

1 Rene Böheim and Mark Taylor, 'Option or Obligation? The Determinants of Labour Supply Preferences in Britain' and 'Actual and Preferred Working Hours', Working Papers 2000-5 and 2001-6, Institute for Social and Economic Research at the University of Essex.

technology helps us organise and co-ordinate meetings with large numbers of people with little effort, and mobiles help us stay in constant touch, warning people of any impending delays or changes to the arrangement. This is supposed to make our lives easy, to reduce stress. But the real effect is apparently the reverse. We end up being caged by the tools we created to help us. Carl Honoré writes about this in his book, *In Praise of Slow*:[2]

> Right from the start . . . timekeeping proved to be a double-edged sword. On the upside, scheduling can make anyone, from peasant farmer to software engineer, more efficient. Yet as soon as we start to parcel up time, the tables turn, and time takes over. We become slaves to the schedule. Schedules give us deadlines, and deadlines, by their very nature, give us a reason to rush. As an Italian proverb puts it: Man measures time, and time measures man.

A result of this is that we lose the opportunity for time for reflection, and consequently end up living more superficial lives. Reflection on decisions, our relationships with people or the world around us go out of the window. As Milan Kundera writes in his novel, *Slowness*, 'When things happen too fast, nobody can be certain about anything at all, not even about himself.'[3] Time for reflection isn't wasted time; it's critically important for our well-being. It's time on the sofa just thinking, time out walking, or it's the five minutes or so spent waiting for the bus when there's nothing else to do — and you don't just get your mobile out and start texting. During reflective time we are turned in on ourselves. It's a state of mind where unexpected thoughts can well up from the subconscious and surprise and delight us.

We end up being caged by the tools we created to help us

Here lies another paradox. Our technological tools are supposed to make time for us, not take it away. While the Earth Simulator Centre is performing its 36 trillion calculations per second, the person with the calculator should be saving 2.1 million years' worth

2   C. Honoré, *In Praise of Slow* (London: Orion, 2004): 21.
3   M. Kundera, *Slowness* (London: Faber & Faber, 1994).

of button pressing. More than 200 years ago Benjamin Franklin envisioned a future in which new machines would make a 4-hour working day enough to provide a comfortable living for all. In the mid-20th century new digital technology seemed to promise a future of leisure for humankind. There were even concerns about how to fill it all.

So where is all this free time? Ask people and most will say they have less free time nowadays, not more. Total working time declined in the last century, but the working day, and the time we spend thinking about work, is stretching out into the mornings and evenings, threatening to completely envelop our waking hours. And, just as advertisements fill up space everywhere we look, opportunities for consumption fill up every moment.

> We lose the opportunity for time for reflection, and consequently end up living more superficial lives

Howard Rheingold wrote in his book, *Smart Mobs*,[4] that:

> Time is socially perceived as something that must be filled up to the very smallest folds, thus eliminating the positive aspects of lost time that could also fill up with reflection, possible adventures, observing events, reducing the uniformity of our existence and so on.

With wonderful efficiency, new digital technology helps us fill up lost time with opportunities to consume unsustainably, to earn more money and to spend more money. It often seems as if we have lost our humanity and have become cogs in a giant, planet-encompassing, wealth-generating, self-consuming machine. This is a point made by Peter Heintel, the Austrian philosopher and founder of the Association for the Deceleration of Time. He writes that 'by living this high-speed hyper-efficient life we lose what makes us humans'.[5]

---

4 H. Rheingold, *Smart Mobs: The Next Social Revolution* (Cambridge, MA: Perseus, 2002).
5 P. Heintel, 'Spänn av', interview in *Dagens Nyheter* (Sweden), 29 November 1996.

Machine time

How did this happen? It's a story the origins of which lie in the clock towers that spread across Europe in the 13th and 14th centuries. The bell would ring out regularly and could be heard, in a world with less noise pollution than today's, for miles in every direction. The distance the bell could be heard was often used to define the limits of a community's jurisdiction. People working in the fields would know the time by the ringing of the bell, not just by the position of the sun, the quality of the light or the activity of the plants and animals around them. Their day for the first time could be ordered mechanically. Down the centuries, clock time spread into almost every sphere of human activity, from schools and hospitals to prisons and workplaces.

Clock time had its apotheosis in the factories of the industrial age. Early in the Industrial Revolution, machines were invented that were more effective at performing simple, repetitive tasks than people were. Production moved away from people's homes and small workshops and into factories, which could accommodate the great size of the new machines and generate efficiencies of scale. Workers were called to the factory at a certain time in the morning, would clock in, work the requisite number of hours, clock out and return home. Their houses were built around the factory and beneath the factory clock. The inhabitants lived lives precisely regimented by the needs of the machines they operated.

The machines of the factory could work continually, unlike humans. Any periods of rest were wasted time for the factory owners. And so working patterns such as shifts were designed to fit people into machine time and maintain production night and day. Production processes were analysed and timed to reach optimal efficiency and rationalisation. F.W. Taylor's time study cut the cost of handling materials at Bethlehem Steel in the US by half, reducing the number of shovellers from 600 to 140. By the end of the 19th century Taylorism had increased the pace of production in factories dramatically.

Eternally Yours[6]

Ever wondered why your car keeps breaking down, or your computer needs replacing after six months? In these days of rapid technological advance, product life and design cycles are becoming ever shorter. New ranges and styles with updated features or functionality are launched dizzyingly fast. Products are now purchased, used and discarded at breakneck speed. A new must-have PDA that doubles as a camera and programmes your dishwasher is sure to come onto the market every month. Recent research from mobile telecoms consultancy Mobile Youth found that 700,000 (20%) of primary school children in the UK own mobile phones. Innovation fuels a fast-paced economy and feeds our spiralling lust for material gratification: according to Laurie Anderson, it 'has turned into a bad habit instead of a way to improve things'.[7] The result? A growing mountain of waste, increased resource and energy use and a 'throw-away culture' that values little but consumes much. Something needs to change.

In 1996 an international team of designers and thinkers based in the Netherlands, inspired by sustainable design guru Enzo Manzini and led by designer Liesbeth Bonekamp and publicist Ed van Hinte, founded Eternally Yours to do just that. This pioneering venture looks at how to increase the durability (product lifetime) of consumer goods, and in the process contribute to sustainable development by reducing waste, pollution, energy and resource use, and even improve our relationship with the objects around us. The logic is simple, the answers less so: extending the life of a product has both complex material and psychological implications. Eternally Yours projects — conducted by institutions such as the Design Academy, Eindhoven; MIT; the University of Helsinki Department of Cultural Studies; and the Department of Philosophy at the University of Twente — examine both aspects of the problem.

Products tend to be discarded for one of three reasons: technical, stylistic or physical obsolescence. Technical obsolescence occurs when a new type of product has a function that supersedes the old design, such as DVDs overtaking videos, for example. Style obso-

6 www.eternally-yours.org
7 In E. van Hinte (ed.), *Eternally Yours: Visions on Product Endurance* (010 Publishers, 1997).

lescence is another word for fashion: last season's clothes are no longer desirable when next season's clothes hit the shops. And, finally, a product might be discarded simply because it doesn't work any more, as when a vacuum cleaner breaks down: this is physical obsolescence. Many of today's products don't have it in them to become old. First they are new, then second-hand, then worthless. Research shows that about 25% of vacuum cleaners, 60% of stereos and a staggering 90% of computers still work when people get rid of them. We deplore the lack of quality that means we won't be able to hand down objects of worth to future generations as our grand-parents or great-grandparents did, but we take it for granted.

But time *can* be incorporated into product design, as companies well know, since obsolescence is often deliberately built into many goods. In the 1950s and 1960s, car manufacturers famously delayed the introduction of aluminium technology, which would have made cars lighter and less prone to rust, hence longer-lasting, thus forcing people to buy new cars more often. If designers can make an item less likely to last over time, then the opposite is also true. Durability can mean making a product more hard-wearing, upgradeable, service-able, recyclable or designing it so that it can be shared, even through successive generations. One example — from the Eternally Yours project — is the table designed by Monique Gerner which has rubber hooks on which children can hang toys or adults can store kitchen towels, meaning that one individual could use it throughout his or her entire life.

The first Eternally Yours conference in 1997 led to the publication of a design sustainability bible, *Visions on Product Endurance*, avail-able as a free download from the website. A second conference in 2003 on Time in Design gave many examples of ways to prolong product life-span. 'Design for sustainability means fostering innova-tion, not just in products and services, but in work methods, behav-iours, and in business processes,' keynote speaker John Thackara explains. Projects completed under the aegis of the Foundation include a number of pioneering designs, such as Nicole van Nes's 'Teletangram', a basic telecommunications device which starts out as a phone, and then acquires further functions such as fax, answering machine, or baby alarm according to the user's needs and desires. Compared with buying new versions or separate items, upgradeable products such as this one have the scope to reduce environmental impact by a third. Pieter Desmet developed an emotion-measuring instrument in order to study the factors that determine someone's choice of mobile phone, to help designers understand what it takes for something to have 'psychological durability', while 'Proud Plastics' investigated the way people value plastic objects over time.

'Value is not limited to product materiality', claims the Foundation. 'Time in design revolves [around] many subjects: fashion, ageing, wear, communication, rootedness, rituals, materials, evolution and time itself.' Eternally Yours asks important questions about what we make and buy, and in hopes of bringing about change that would be both ecological and sociological, to countering what it labels as 'enormous waste and needless destruction of value'. It does not (and perhaps cannot) provide definitive answers, but it has begun to make an impact at the forefront of product innovation; and, if it can spread its message of long-term sustainability, it may help change the face of design forever.

Chris Sherwin and Mireille Kaiser

Although many of the products we consume are produced in just such a fashion in developing countries, today in the developed world the factory system no longer dominates. But we are still socialised into the metronomic rhythms of Taylor's efficiency scheme.

Machine time doesn't replace natural time, but slides along next to it. We notice the return of the seasons, the sun in the morning, the rotation of the night sky, seemingly unchanging and never-ending cycles. We are born, have children, grow old and die, and see the same happening to others. Meanwhile, we are acutely aware of the regular beat of the machine, marking out **To separate us from** linear time, transporting us with exhilarating **natural time is** speed from the past into the future. The edge of **dehumanising** the future is our favourite place to be: we live on the cusp of the next deadline, of the next holiday, the next general election, the next announcement of financial results, always on the way to the next event.

This is where the abstraction of time takes over from the experience of time, when machine time dominates natural time. The philosopher Martin Heidegger is one of the many critics of machine time. He argued that it distorts our relationship with reality, as time becomes an object or a quality of objects instead of something intrinsic to what we are and what we do. To separate us from natural time is dehumanising.

And the beat is getting faster — impossibly fast. If the way we live our lives has so much to do with machine time, then digital technologies and the boom in computing power are speeding up our lives unsustainably, by adding noughts to the end of what machines can do.

Timeless time

It's not surprising that so many people want to opt out, downshift, get out of the rat race. The (mostly broadsheet) newspapers in the

UK feature articles every month or so about people leaving the city, and finding a rural retreat in the Alpujarra or the Outer Hebrides, somewhere they can reconnect with nature and 'find themselves'. And, for every person or family trying to leave machine time behind them, there's a movement aimed at slowing society down, or reconfiguring our relationship with time in some way.

Peter Heintel's association is one of those. The Association for the Deceleration of Time has around 1,000 members across Europe. The members all share the view that we are losing our quality of life through the cult of speed and the need to quantify progress. They meet once a year and indulge in decelerating activities, such as a 100-metre slow race, in which the competitors must cover the distance in as much time as possible. Slowness rather than speed, patience rather than power, are required in order to win.

If enough people join clubs like Peter Heintel's, the Slow Food Movement, or the Sloth Club of Japan, which advocates a less hurried and greener lifestyle, then they might have a positive effect on a society-wide scale. And there's a great deal that government policies and sympathetic employers can do to help people gain more control of their own time, from supplying guidance on work–life balance to longer parental leave. But let's go back to digital technology. It has provided us with organisational tools that seem to take control of time away from us, and digital products that are designed to make what we do more efficient, but make us feel more busy and stressed. At the same time as speeding the quickening beat of the machine, these information and communication technologies (ICT) could be changing our relationship with time in perhaps unexpected ways, and actually loosening the grip of machine time.

> The Association for the Deceleration of Time meet once a year and indulge in decelerating activities, such as a 100-metre slow race

Let's look at a couple of examples. First, the World Wide Web.

Physically, the World Wide Web is made up of millions of computers all linked directly or indirectly. The computers are located everywhere around the world, from purpose-built server farms in India to the corner of someone's bedroom in California.

Together, all these computers store a staggeringly vast amount of information. Every page of information has a unique address and can be accessed by anyone from anywhere. There is so much information that we have trouble coping with it and need to use search engines to navigate it. Google, the most popular search engine, said it was performing over 350 million searches a day in 2005, on over 8 billion pages of material.

The web provides us with theoretically instantaneous retrieval of information — though we may have to wait a few seconds for the pages to load. All of history is recorded on the web as it is made, and will exist there in perpetuity. Even if the original files are deleted, it is likely that somewhere on the web the information will continue to exist, as the files will have been cached countless numbers of times by different search engines, or copied onto individuals' computer hard drives.

The history is there as long as the web and computers exist, but in no noticeable order. It certainly isn't linear. You can't (as yet) order web pages according to when they were created, or move about the web chronologically, and it wouldn't make a lot of sense to do that even if you could. Time on the web is *timeless time*, to borrow the phrase of the sociologist Manuel Castells. Events become locations in the dispersed web, joined together by hyperlinks which transport the user from page to related page with minimal effort. The events have no temporal relationship to each other: they are locations we can visit whenever we want, rather than fixed points in past time that can't be returned to.

Mobile space

Mobile telephones change our relationship with time in a different way.

The Japanese academics Mizuko Ito and Tomoko Kawamura have researched young people's use of mobiles.[8] As part of her research

---

8   Conversation between Howard Rheingold, Mizuko Ito and Tomoko Kawamura, Tokyo, October 2001; quoted in H. Rheingold, *op. cit.*

she observed a group of 30 young people in Japan organising an evening of karaoke using their mobile phones: 'As the date grew nearer, the frequency of messages increased. But only four people showed up on time at the agreed place.' There were no recriminations for others turning up late or missing the appointment completely, as most people stayed in touch with their mobile phone, texting and leaving voice messages. Kawamura concluded that for young people today it is perfectly acceptable to show up late or not show up at all as long as they are present in the same mobile communications space.

This mobile space is different from the vast formless space created by the World Wide Web. It is only open to trusted intimates, members of the group. It forms an open channel of contact in which communication between different people and groups can continue without restriction. It is unbounded physically because it doesn't exist in the physical world, and it is unbounded temporally because text and voice messages can be left at any time and picked up later. Your mobile can be switched off for a while and the messages will be there when you turn it back on. You can reconnect with the mobile space when your face-to-face activities allow you to and — importantly — when you want.

> For young people today it is perfectly acceptable to show up late or not show up at all as long as they are present in the same mobile communications space

Research in Norway[9] has shown similar behaviour to that in Japan: people with mobiles make less effort to plan definitely in advance, and care less about being on time for appointments. Esben Tuman Johnsen, from the Norwegian mobile phone company Telenor, writes that 'the opportunity to make decisions on the spot has made young people reluctant to divide their lives into time slots, as older generations are used to doing'. His colleague Rich Ling reports a 'softening of time' among the young around the world who text each other.

9   R. Ling and B. Yttri, 'Hyper-coordination via Mobile Phones in Norway', in M. Aakhus and J. Katz (eds.), *Perpetual Contact: Mobile Communication, Private Talk and Public Performance* (Cambridge, UK: Cambridge University Press, 2002).

The mobile generation is more inclined to live on the hoof and less likely to be bound by formal time – 'clock time' – structures. They form groups of loosely co-ordinated individuals who come together and part as necessary or desired, a kind of mobile phone-co-ordinated public flocking. Commentators have called this phenomenon 'swarming': Kevin Kelly, author and co-founder of *Wired* magazine, wrote that 'What emerges from the collective is not a series of critical individual actions but a multitude of simultaneous actions whose collective pattern is far more important. This is the swarm model.'[10]

The World Wide Web has been around for a decade and mobile phones are an even more recent phenomenon. Yet already they appear to be having an influence over the way we use and think about time

This behaviour is reminiscent of Jay Griffiths's description earlier in this volume (page 55) of time-organisation in non-Western cultures:

> for every indigenous group I have ever known or read about, timing in social interactions is indeterminate, unpredictable, demanding flexibility, fluidity and quick co-ordination. It is a graceful, alert way to live, demanding acute skills of psychology ... Hunter-gatherer time is a series of unique moments, confluences of a hundred streams, a thousand interconnecting factors ... Scheduling or planning would destroy the necessary elusiveness of this subtle sense of timing, and would kill stone dead the exquisite sense that time is alive.

The World Wide Web has been around for a decade and mobile phones are an even more recent phenomenon. Yet already they appear to be having an influence over the way we use and think about time. Could digital technology be reconfiguring our relationship with time, opening up an escape from the tyranny of the clock? Maybe. It will certainly be interesting to watch as today's teenagers, who have grown up with this technology, become parents, employers and politicians. Linear machine time will no doubt

10   K. Kelly, *Out of Control: The New Biology of Machines, Social Systems and the Economic World* (Perseus Books, 1995).

continue speeding into the future, accelerating as technology allows ever greater efficiency and ever greater consumption, leading to greater stress in a 24-hour society. Meanwhile, the timeless time of the web and the loose, flexible time of mobile space will exist and perhaps grow alongside linear machine time and the cycles of natural time. They could, perhaps, provide some respite, allow us a little more control and the opportunity to relax and reflect occasionally.

10

Conclusion

Vidhya Alakeson

Our relationship with time has become distinctly contradictory. We snap up every new gadget that hits the market, hoping that it will allow us to cram just one more thing into our everyday lives. And, while firing off emails on the journey home from work, we dream of 'downshifting' and leaving it all behind. Are our lives best lived in the slow or fast lane? If only someone would tell us what to do about time.

Under the Ceauşescu regime in Romania in the 1980s, each individual in Bucharest was allocated a time (reservation sounds far too glamourous) to dine at one of the 'people's restaurants' in the city. Of the people in Bucharest today who have spoken to me about the scheme, few mention the abominable food served by the so-called restaurants: it was the indignity of being told by the authorities when to eat to which they objected.

A central diktat to speed up or slow down is doomed to failure. It would be impossible to set a pace that would suit everyone. Different people value the opportunity to live their lives in different ways and at different speeds. There will always be workaholics who are unable to tear themselves away from their desks, just as there will always be those who only work to live. And, at different points in our lives, we want the freedom to choose a different pace.

The freedom to choose

This collection of essays has explored many facets of our relationship with time, from cosmological time to working time, nature's time to time politics. A central theme running through many of the essays is the importance of individual choice in determining our relationship with time. Our experience of time is very much conditioned by how much choice we feel we have in the way we spend our days. The chances are that an employee whose boss bombards him with email at five o'clock on a Friday evening feels very differently about the efficiency of the technology from a freelancer who has greater say over when to respond and when to shut down his computer. The freedom to choose whether we live our lives in the fast or the slow lane and the freedom to move from one to the other as our lives change is the real issue underpinning our fraught relationship with time.

> A central diktat to speed up or slow down is doomed to failure

The apparent lack of freedom to choose identified by many of the contributors seems paradoxical in a liberal democracy that cherishes the principle of individual liberty. But choices don't exist in a vacuum; they are context-specific, shaped by the society in which we live. Philosopher John Stuart Mill[1] argued for the protection of individual liberty against 'the despotism of custom' as well as against the tyranny of government:

> There needs protection against the tyranny of prevailing opinion and feeling; against the tendency of society to impose, by means other than civil penalties, its own ideas

---

1 John Stuart Mill, *On Liberty* (London, 1859).

and practices as rules of conduct on those who dissent from them.

In *A Better Choice of Choice*,[2] sustainable development commentator Roger Levett illustrates the often illusory nature of choice that we enjoy. He gives the example of transport. In theory, we can choose to cycle rather than travel by car or bus. But the choices of millions of other commuters to take to their cars makes cycling an unpleasant, if not dangerous, way of getting around and our choice is, therefore, constrained.

Our problematic relationship with time stems from a similarly constrained set of choices. We might kid ourselves that we can freely choose our own pace but society's values and expectations constrain us and flouting expectations requires courage and determination. Much of the stress and anxiety that is consistently reported when people are asked about their relationship with time stems from this lack of meaningful choice.

> We might kid ourselves that we can freely choose our own pace but society's values and expectations constrain us

The experience of many new fathers is a good example. Only one in five fathers in the UK takes the full two weeks' paid paternity leave to which he is legally entitled.[3] Even in Sweden, the world leader in generous family-friendly policies, only 35% of fathers take their full six months' entitlement. Many more say they would do, in theory. Two-thirds of working men in the UK say that they would be likely to take paternity leave and 26% are certain to take time off.[4] For many families, the financial reality of having a child prevents fathers staying away from work as long as they would like to. But, as writer Richard Reeves argues in an essay on the new politics of gender, 'It is not just that most breadwinners are men, it is that, to

2   R. Levett, *A Better Choice of Choice: Quality of Life, Consumption and Economic Growth* (London: Fabian Society, 2003).
3   DTI (Department of Trade and Industry) figures, 2004.
4   MORI, *Men Support Blair's Paternity Leave* (2000).

be a man, you have to be a breadwinner.'[5] As long as society's view of men is bound up with going to work, men will not be entirely free to choose to stay at home.

Only one in five fathers in the UK takes the full two weeks' paid paternity leave to which he is legally entitled

Constrained choices

Western, consumerist society has a very clear narrative about what is a valuable way to spend time; which activities are meaningful and which aren't; whose time is valuable and whose can be wasted without a second thought. Jay Griffiths explains in her chapter how Christianity, then colonialism and now globalisation have sought to impose these values on other cultures, removing the freedom of indigenous peoples to live according to their own rhythm and values.

I noticed recently that the stock response to 'how are you?' is no longer 'fine, thank you', but 'busy'. Being busy at work is a sign of success and being successful in life has become closely tied to success in paid employment. We live in a society that deifies paid work, as if paid work were the only route to happiness. While it is undeniably true that meaningful employment is vital for life satisfaction, it is equally the case that increases in income are not matched by quality-of-life improvements. While household income in the UK has risen substantially since the early 1970s, the proportion of those who are fairly or very satisfied has risen only marginally and those who are not very or not at all satisfied has similarly fallen only slightly.[6] Expectations

I noticed recently that the stock response to 'how are you?' is no longer 'fine, thank you', but 'busy'

5   R. Reeves, 'Men Stuck in Cages of their Own Creation', *New Statesman*, 16 August 2004.
6   N. Donovan and D. Halpern, *Life Satisfaction: The State of Knowledge and Implications for Government* (London: UK Cabinet Office Strategy Unit, 2002).

tend to level up, leaving us feeling no better off despite an increase in absolute income. But that doesn't stop us working the longest hours in Europe and taking the shortest holidays. Paid work has become our main source of identity and worth as well as our main source of income.

One of the consequences of the idolatry of paid work is that the massive contribution of parents, grandparents and other unpaid carers goes completely unrecognised and unrewarded. The Office of National Statistics estimates that this unpaid care in the UK is worth a staggering £929 billion a year, or 104% of GDP. But lack of recognition makes it more difficult for people to choose to spend their time fulfilling these vital caring roles. In recent years researchers have identified the importance of a caring, stimulating home environment for the positive cognitive and behavioural development of children. This has placed greater pressure on parents to become more involved in their children's earliest years. But, as a society, it's not the women who leave their careers behind and choose to stay at home who win our admiration. It's Cherie Blair and other superwomen like her who remain at the peak of their careers at the same time as having happy, healthy children.

We have an equally clear idea of whose time, and related to that whose knowledge, is valuable. An hour of footballer David Beckham's time would set you back tens of thousands; an hour with the best doctor or lawyer several thousand. An hour of a hospital porter's or a school dinner lady's time would cost you in the region of five pounds. It's not a simple case of economics – there's only one David Beckham so you have to pay a huge fee for his unique skills. Societal values are bound up with the economics to determine how much we're willing to pay for someone's time. As Polly Toynbee vividly describes in her account of life on the minimum wage, the poor are kept waiting as if their time had no value when, in actual fact, time really is money for low income earners who are just scraping by.[7]

7   P. Toynbee, *Hard Work* (London: Bloomsbury, 2003).

> I sat there thinking how low value permeates everything about the lives of the poor. I had queued in the post office to pay the rent, trekked to the only shop that recharges the electric meter and queued again. Everywhere I was kept waiting and yet my time was precious because, like all the people at the [job] agencies, I needed to get a job quickly. But poor people's time is regarded as valueless. At £4.10 an hour, it is almost.

In his essay, David Boyle explains how an essential part of the success of time banks has been to undermine accepted notions of whose time is valuable. Within a time bank, everyone has something to offer and can be part of an ongoing network of reciprocal relationships. As he rightly observes, the effect that feeling useful can have on people who have spent most of their lives being categorised as disabled or useless can be transformative.

The answer to our uneasy relationship with time lies not in determining the pace at which people should live their lives but in creating an environment that allows more people to exercise the freedom to choose in a meaningful way. As Nobel Prize-winning economist Amartya Sen argues,[8] the ultimate goal of progress should not be to increase income but to increase the substantive freedoms available to every individual to choose a life that he or she finds of value. The challenge we face is to reframe the choice sets on offer so that people have the freedom to choose a meaningful life, whether that is lived in the slow or fast lane. This means changing the values and expectations that shape our experience of time.

As a society, it's not the women who leave their careers behind and choose to stay at home who win our admiration

8   A. Sen, *Development as Freedom* (Oxford, UK: OUP, 2001).

Choice reframed

The right of parents of young children to demand flexible working hours is a recent example of a government policy that seeks to reframe the choices that parents face between work and family life, and early indications are that it has been successful in giving parents a meaningful choice. A range of other proposals that offer parents more choice will be implemented in the government's third term: for example, the extension of paid maternity leave to nine months, with greater entitlements for fathers. In his essay, Geoff Mulgan identifies the extension of such entitlements as central to improving the quality of working life in an economy where the drive for human capital improvement is relentless.

But more radical proposals may eventually be necessary to significantly reshape our relationship with time. Building periods of community service formally into the curriculum for secondary and tertiary education could raise the status of non-paid work. Sabbaticals, offering people the opportunity to try out new activities, develop new skills, relax and have fun, could create new sources of personal identity and worth for today's working population, ending our single-minded fixation with work as a source of meaning.

Government can provide some of the architecture for the kind of culture change we require. But other actors in society have an equally important contribution to make. The centrality of the tussle between work and family life to our relationship with time highlights the importance of employers in enabling meaningful choices about how people spend their time. Will Hutton and Alexandra Jones conclude that there needs to be a change of mindset in organisations away from thinking that time is money towards the concept of quality time. Organisations need to create the right incentives to secure productive time from their employees, not just long hours.

Opinion is divided on the role that new technologies have to play. For Jonathon Porritt, technology is implicated in the unsustainable acceleration of society. It oils the wheels of commerce that drive a society fixated on consumption and material progress. In contrast,

James Goodman and Britt Jorgensen foresee a more optimistic technological future. New technologies could also play a positive role in reshaping the choices we face, just as they have started to change the way we communicate and think about time. The Internet and mobile communications undoubtedly offer tremendous potential to underpin the flexible working patterns that will promote a better balance between work and family commitments. Whether this potential is realised depends less on the technologies themselves and more on the policy and organisational frameworks that underpin them. There is no simple techno-fix to our relationship with time.

Securing choice for the future

Biologist Ghillean Prance beautifully describes the precision of nature's clock that sets flowers to bloom just as insects arrive to pollinate them and regulates the arrival of predators and prey to keep ecosystems perfectly balanced. It is precision timing that has developed over millennia and is now severely under threat from human activity. Human beings are disrupting nature's fine balance faster than nature can keep up, threatening species with extinction and ultimately destroying the ecosystems on which human life depends. In fact, our current rate of exploitation of the Earth threatens not only the survival of our own species but also the post-human future of the planet, stretching for several billion years. As Sir Martin Rees warns, there is a real danger that human activity will curtail the planet's entire potential.

For much of the modern era, securing meaningful choices in the present has been at the expense of the future

Planet Earth bears the imprint of the billions of years of evolution and simultaneously holds the rich promise of the future. But we human beings too often lack the imagination to situate the present in a time continuum that stretches from the past to the far future.

For much of the modern era, securing meaningful choices in the present has been at the expense of the future. Left unchecked, current levels of environmental destruction will preclude future generations from many meaningful choices such as the ability to live free of pollution, not to mention the plant and animal species that will only be pictures in books for tomorrow's children. We act as if there were some uncertainty over whether or not the future will ever arrive and foolishly ignore the consequences of present actions. Our reluctance to acknowledge our own mortality blinds us to the continual passing of time and arrival of the future.

Our presence on the planet may be fleeting but the value of freedom of choice is timeless. Previous generations fought against dictatorial regimes and oppressive social conventions to secure the freedom we enjoy in the present. We have a responsibility to pursue meaningful choices for ourselves while securing the same freedom for future generations. Time — past, present and future — is the very essence of sustainable development. Building a more positive relationship with time in which individuals set their own pace is at the heart of a more sustainable future.

Contributors

Vidhya Alakeson

Born in 1976, Vidhya Alakeson is a research fellow at the Social Market Foundation. She was formerly a principal policy advisor at Forum for the Future where she ran the Digital Europe project, and has worked at Policy Network and the Foreign Policy Centre. She is one of the co-authors of *Making the Net Work: Sustainable Development in a Digital Society* (Xeris Publishing, 2003) and co-author of *Going Public: Diplomacy for the Information Society* with Mark Leonard (The Foreign Policy Centre, 2000).

Tim Aldrich

Born in 1975, Tim is an executive advisor at KPMG. He edited this book while working as a senior advisor at Forum for the Future, specialising in communications and technology. He joined Forum from the RSA where he worked on the *Journal*. He is one of the co-authors of *Making the Net Work: Sustainable Development in a Digital Society*.

David Boyle

Born in 1958, David Boyle is a writer and journalist focusing on new ideas around the environment, politics, economics and the future. An associate at the New Economics Foundation, he has edited a number of magazines and journals and is the author of *Funny Money: In Search of Alternative Cash* (HarperCollins, 1999); *The Tyranny of Numbers: Why Counting Can't Make Us Happy* (HarperCollins, 2000); and *Authenticity: Brands, Fakes, Spin and the Lust for Real Life* (HarperCollins, 2004).

James Goodman

Born in 1972, James Goodman is a principal advisor in the Forum Business Programme. He manages partnerships with a number of Forum for the Future's business partners and specialises in the relationship between technology and sustainable development. James's publications include *Making the Net Work: Sustainable Development in a Digital Society*, a book based on research for the European Commission-funded Digital Europe project, and *The Future Impact of ICT on Environmental Sustainability*, based on foresight work for the Institute for Prospective Technology Studies.

Jay Griffiths

Jay Griffiths is a writer based in Wales and the author of *Pip Pip: A Sideways Look at Time* which, republished in the US as *A Sideways Look at Time*, won the Barnes & Noble Discover Great New Writers™ award for non-fiction 2003. She has written for a number of publications including *New Internationalist* and *The Ecologist*. Jay Griffiths's next book deals with the idea of wilderness.

Will Hutton

Will Hutton is chief executive of The Work Foundation and columnist for *The Observer*, of which he was editor-in-chief for four years in the 1990s. Will started his career as a stockbroker before moving into radio and television journalism at the BBC, including a time as economics correspondent for Newsnight in the mid-1980s. He has written several bestselling books including *The State We're In: Why Britain Is in Crisis and How to Overcome It* (Vintage Books, 1996); *The State to Come* (Vintage Books, 1997); and *The World We're In* (Abacus, 2003).

Alexandra Jones

Alexandra Jones is a senior researcher at The Work Foundation, where her work focuses on flexibility, diversity, workforce development and public services. She also runs the Employers for Work–Life Balance website and is one of the founding members of the Public Services Unit at The Work Foundation. Previously Alexandra worked as a private secretary for the Permanent Secretary at the Department for Education and Skills and as a researcher at the Institute for Public Policy Research.

Britt Jorgensen

Born in Denmark in 1973, Britt Jorgensen works at OneWorld International as project manager on the Open Knowledge Network. Previously, she has worked at the Danish think-tank Mandag Morgen and was a researcher at Forum for the Future on the Digital Europe project. She is one of the co-authors of *Making the Net Work: Sustainable Development in a Digital Society*.

Mireille Kaiser

Born in Boston, Massachusetts, in 1981, Mireille Kaiser is currently studying for a master's degree in social anthropology at King's College, Cambridge. She was educated in the US and France before completing her BA in English and modern languages at Lady Margaret Hall, Oxford, in 2002.

Geoff Mulgan

Geoff Mulgan is director of the Institute of Community Studies, having been Head of Policy at 10 Downing Street and before that director of the Cabinet Office's Performance and Innovation Unit. In 1993 he founded the independent think-tank Demos and was its director until 1997. Geoff Mulgan has published on a wide number of policy areas and is the author of *Connexity: How to Live in a Connected World* (Chatto & Windus, 1997). He was awarded a CBE in 2004.

Jonathon Porritt

Born in 1950, Jonathon Porritt is programme director of Forum for the Future and chair of the UK's Sustainable Development Commission. Initially working as a teacher in London during the 1970s, he served as chairperson of the Green Party during 1979-80 and 1982-84, giving up teaching in 1984 to become director of Friends of the Earth. In 2003-04 he sat on the Royal Society/Royal Academy of Engineering Working Group on Nanotechnology. Jonathon Porritt has written extensively on the environment, science and sustainable development including *Seeing Green: The Politics of Ecology Explained* (Blackwell, 1984) and *Playing Safe: Science and the Environment* (Thames & Hudson 2000). He was awarded a CBE in 2000.

Ghillean Prance

Born in 1937, Sir Ghillean Prance is scientific director of the Eden Project in Cornwall, visiting professor at Reading University and McBryde Professor at the US National Tropical Botanical Garden in Hawaii. He was director of the Royal Botanic Gardens, Kew, from 1988 to 1999. Ghillean Prance has held positions in the US and Brazil and has been widely honoured for his work on the ecology of Amazonia. He was knighted in July 1995, and received the Victoria Medal of Honour from the Royal Horticultural Society in 1999.

Martin Rees

Born in 1942, Sir Martin Rees is the Astronomer Royal, Professor of Cosmology and Astrophysics at Cambridge University and Master of Trinity College, Cambridge. One of the world's leading astronomers, he has authored or co-authored 500 research papers. Martin Rees is also a well-known commentator in public on science and public policy. His most recent book, one of five for the general reader, is *Our Final Century? Will the Human Race Survive the Twenty-first Century?* (Random House, 2003) He was knighted in 1992.

Chris Sherwin

Born in 1969, Chris Sherwin is a principal sustainability advisor at Forum for the Future, having joined in 2004 from Philips Electronics, working on eco-design and sustainable innovation. He has lectured and published extensively on sustainability and design both in the UK and abroad.

Mary Warnock

Born in 1924, Baroness Mary Warnock is one of the world's leading ethicists and educationalists. Best known for chairing the Committee of Inquiry into Human Fertilisation and Embryology (1982-84) and writing the subsequent 'Warnock' Report, she has been a tutor in philosophy at St Hugh's College, Oxford, Headmistress of Oxford High School and Mistress of Girton College, Cambridge. Awarded a life peerage in 1985, she sits as an independent peer in the House of Lords and her writings include *Existentialism* (Oxford University Press, 1970); *Imagination* (University of California Press, 1976); *Memory* (Faber & Faber, 1987); and *An Intelligent Person's Guide to Ethics* (Duckworth, 1998).